Information Circular 9519

Mine Roof Bolting Machine Safety: Investigations of Roof Bolter Boom Swing Velocity

By Joseph H. DuCarme and August J. Kwitowski

DEPARTMENT OF HEALTH AND HUMAN SERVICES
Centers for Disease Control and Prevention
National Institute for Occupational Safety and Health
Office of Mine Safety and Health Research
Pittsburgh, PA • Spokane, WA

February 2010

This document is in the public domain and may be freely copied or reprinted.

Disclaimer

Mention of any company or product does not constitute endorsement by the National Institute for Occupational Safety and Health (NIOSH). In addition, citations to Web sites external to NIOSH do not constitute NIOSH endorsement of the sponsoring organizations or their programs or products. Furthermore, NIOSH is not responsible for the content of these Web sites. All Web addresses referenced in this document were accessible as of the publication date.

Ordering Information

To receive documents or other information about occupational safety and health topics, contact NIOSH at

Telephone: **1–800–CDC–INFO** (1–800–232–4636)
TTY: 1–888–232–6348
e-mail: cdcinfo@cdc.gov

or visit the NIOSH Web site at **www.cdc.gov/niosh**.

For a monthly update on news at NIOSH, subscribe to NIOSH *eNews* by visiting **www.cdc.gov/niosh/eNews**.

DHHS (NIOSH) Publication No. 2010–126

February 2010

SAFER • HEALTHIER • PEOPLE™

CONTENTS

Page

Abstract 1
Introduction 2
Background 3
Phase 1 tests: NIOSH volunteer subjects 5
    Study population 5
    Experimental design and measurements 6
Data analysis: NIOSH subjects 10
    Methods 10
    Results 10
    Discussion 19
Phase 2 tests: mine worker subjects 19
    Study population 19
    Experimental design and measurements 20
Data analysis: mine worker subjects 23
    Methods 23
    Results 25
    Discussion 28
Conclusions 28
Acknowledgments 29
References 29
Appendix.—Mine worker data 31

ILLUSTRATIONS

1. Typical roof bolter posture 4
2. Overhead view of bolting environment 5
3. Slide mechanism used for NIOSH subject testing 6
4. Standing posture 7
5. Stooping posture 7
6. Squatting posture 7
7. Kneeling posture 7
8. JACK markers, front 8
9. JACK markers, back 8
10. NIOSH subject test fixture and artificial roof 9
11. Velocities for forward boom swing motions 12
12. Velocities for backward boom swing motions 12
13. NIOSH subjects: 72-in seam height, standing posture, forward motion 14
14. NIOSH subjects: 72-in seam height, standing posture, backward motion 14
15. NIOSH subjects: 60-in seam height, stooping posture, forward motion 15
16. NIOSH subjects: 60-in seam height, stooping posture, backward motion 15
17. NIOSH subjects: 60-in seam height, squatting posture, forward motion 16
18. NIOSH subjects: 60-in seam height, squatting posture, backward motion 16

CONTENTS—Continued

Page

19. NIOSH subjects: 48-in seam height, squatting posture, forward motion 17
20. NIOSH subjects: 48-in seam height, squatting posture, backward motion 17
21. NIOSH subjects: 48-in seam height, kneeling posture, forward motion 18
22. NIOSH subjects: 48-in seam height, kneeling posture, backward motion 18
23. Mine worker subject in a stooping posture 21
24. Mine worker subject in a kneeling posture 21
25. Distance versus time plot for swing-out 24
26. Distance versus time plot for swing-in 24
27. Mine worker subjects: standing 72-in results 25
28. Mine worker subjects: stooping 60-in results 25
29. Mine worker subjects: squatting 60-in results 26
30. Mine worker subjects: squatting 48-in results 26
31. Mine worker subjects: kneeling 48-in results 27
A-1. Actual mine worker data for standing posture, 72-in height 31
A-2. Actual mine worker data for stooping posture, 60-in height 32
A-3. Actual mine worker data for squatting posture, 60-in height 33
A-4. Actual mine worker data for squatting posture, 48-in height 34
A-5. Actual mine worker data for kneeling posture, 48-in height 35

TABLES

1. Detailed results of NIOSH subject testing 11
2. Mine worker subject anthropometric data 20
3. Target swing velocities and times 22
A-1. Results for standing posture, 72-in height 31
A-2. Results for stooping posture, 60-in height 32
A-3. Results for squatting posture, 60-in height 33
A-4. Results for squatting posture, 48-in height 34
A-5. Results for kneeling posture, 48-in height 35

ACRONYMS AND ABBREVIATIONS USED IN THIS REPORT

ANSI	American National Standards Institute
ASIS	anterior superior iliac spine
DOE	U.S. Department of Energy
MSHA	Mine Safety and Health Administration
NIOSH	National Institute for Occupational Safety and Health
PVC	polyvinyl chloride
SD	standard deviation

UNIT OF MEASURE ABBREVIATIONS USED IN THIS REPORT

ft	foot
in	inch
in/sec	inch per second
lb	pound
sec	second

MINE ROOF BOLTING MACHINE SAFETY: INVESTIGATIONS OF ROOF BOLTER BOOM SWING VELOCITY

By Joseph H. DuCarme[1] and August J. Kwitowski[2]

ABSTRACT

An analysis of accident/injury data for 2001 through 2005 from the Mine Safety and Health Administration (MSHA) revealed that powered machinery accounted for nearly 40% of the total underground coal injuries reported and 62% of all fatalities. Underground coal miners work in an environment with limited space for lateral movement and in awkward postures, including kneeling on one or both knees. During informal discussions, MSHA and the United Mine Workers of America expressed concerns about the velocity of appendages on machines used in such environments.

This report describes a study of operator movement relative to the motion of a roof bolting machine boom arm. This work was aimed at reducing the risk of injury to underground coal mine workers from moving machinery. The study used motion capture technology to evaluate human movement in restricted heights and postures while controlling a mockup of a roof bolter boom.

Results suggest that boom horizontal swing velocity is an important factor in determining operator safety from pinch point and crush hazards during the boom positioning phase of the bolting sequence. The working height where the machine is operating, the operator's working posture, and the direction of the swing, toward or away from the operator, are also important in determining safe boom velocity.

---

[1]Mechanical engineer.
[2]Civil engineer.
Office of Mine Safety and Health Research, National Institute for Occupational Safety and Health, Pittsburgh, PA.

INTRODUCTION

An analysis of the Mine Safety and Health Administration's (MSHA) accident, injury, illness database for 2003–2007 shows there were 405 injuries involving worker-to-roof bolter contact. During those 5 years, roof bolting was the most hazardous machine-related job in underground mining. It accounted for nearly one-third of accidents involving powered machinery.

Literature searches showed that worker activities involving boom arm movement during the bolting cycle [Klishis et al. 1993a,b] were associated with many of the accidents. A committee established by MSHA in 1993 composed of representatives from the West Virginia Board of Coal Mine Health and Safety, the U.S. Bureau of Mines, and roof bolter manufacturers identified 10 problems that may have contributed to or caused accidents. Seven of these were associated with movement of the boom arm. MSHA [1994] and Turin et al. [1995] revealed that there are no data on safe velocities for machine appendages operating close to workers in the confined work environment of underground mines. A prior NIOSH study investigated vertical boom velocity [Ambrose et al. 2005; Bartels et al. 2003]. During the course of that research, MSHA, bolting machine manufacturers, and all of the mine worker human subject participants in the study inquired as to when horizontal swing velocity would be studied.

Several studies in the robotics industries have provided data for setting safe machine appendage speeds for reducing injuries and developing numerous guidelines for the safety of workers close to production line robots. Etherton [1987] reported that 10 in/sec is a speed at which humans could recognize and react to a perceived hazard. In addition, the Occupational Safety and Health Administration [OSHA 1987] requires that robot speeds for teach-and-repeat programming sessions where the programmer is within the robot's motion envelope conform to the American National Standards Institute (ANSI) slow speed recommendation. The current ANSI standard recommends that this slow speed should not exceed 10 in/sec. However, Karwowski et al. [1992] reported that test subjects can perceive potential hazards from a moving robot arm at a rate of motion from 8 to 16 in/sec. Their study suggests that the safe speed of robot motions for teaching and programming purposes lies somewhere between 8 and 10 in/sec. Moreover, the U.S. Department of Energy (DOE) recommends a restricted speed of 6 in/sec on any part of the robot when a teacher is within the robot's motion envelope because mistakes in programming can result in unintended movement. This slower speed would reduce possible injuries to a teacher if an inadvertent movement occurred [DOE 1993].

A production line robot will move from point to point regardless of a human in its motion path. Unlike a robot, the motion of the boom arm on a roof bolting machine is under the control of its operator. Observations of operators during bolting operations show that they, when possible, attempt to move in unison with the swing motion of the boom arm. If the motion is too fast for them to follow in the confined environment, they need to release the control so that they may reposition themselves. The designs of most roof bolter controls are quite logical. Pushing the control handle away from the operator causes boom swing in that direction and vice versa. However, if an operator should stumble or slip while swinging the boom out (toward oneself), the natural reaction is to attempt to steady oneself. This could result in unintended motion of the boom arm and injury to the operator.

The goal of this study is to increase the safety of bolting machine operators during horizontal boom swing operations. Boom swing usually occurs when the operator is repositioning the boom arm to a new bolt insertion location. It requires that the operator properly actuate the right control(s) and then reposition his/her body to follow the moving boom arm. In low roof heights, operators may perform this task from a kneeling position, which hinders their ability to keep pace with the boom arm. The basic issue is that it is not known what boom swing velocity maximizes the operator's chances of escaping injuries while still allowing the operator to perform bolting functions in a timely manner. Like the earlier vertical boom arm study, this work used motion capture technology to evaluate human motions while operating a bolting machine in various postures.

Laboratory experiments were conducted in two phases. Phase 1 used NIOSH volunteer subjects to determine the velocities at which they could follow a predetermined path under three working heights. The predetermined path represented the movement required by a roof bolter operator during boom swing operation. In phase 2, experienced mine workers were recruited to measure human motion versus time in operator postures commonly used when operating a roof bolter. The phase 2 study used three boom swing rates: two velocities determined from phase 1 data and the manufacturer's normal swing rate. Analysis of phase 2 data defined the range of roof bolter boom swing velocities that provide the roof bolter operator with the best prospect of safety.

BACKGROUND

Support of the roof after coal extraction is essential to worker safety and ventilation of the mine. Roof bolts are long steel rods, at least 4 ft long, with expansion or resin anchors used to secure them into the roof. After the mining crew removes a section of the coal seam, roof bolting machine operators install these bolts to secure the areas of unsupported roof. The bolters must complete this task as quickly as possible to prevent sections of the roof from falling. The speed at which the bolting crew can work is critical to mine safety. Bolts are installed according to the mine's MSHA-approved roof control plan. Typically, this calls for installation on 4-ft centers. A bolter operator's usual work sequence includes moving the machine into position, setting up for the operation, drilling the hole into the roof, and installing the bolt [Klishis et al. 1993a]. The setup step of this sequence may include setting the automated temporary roof support system, scaling (manual removal of loose material from the roof and walls), handling ventilation material, performing a methane check, emptying the machine's dust box, and other tasks. Drilling bolt holes involves inserting the drill steel in the chuck, adding extension steels if required, changing the bits, drilling the hole, and removing the steel. The bolt installation is accomplished by making up bolt assemblies, inserting resin cartridge into the bolt hole (if used), inserting the bolt into the hole and bending it if necessary, and spinning to mix resin or torque the installed bolt. The bolting sequence repeats until the unsupported area of the roof is secured and the requirements of the roof control plan are met. Then the operator moves the machine to a new location and begins the process again.

Roof bolting is a fairly structured and repetitive process. Figure 1 shows an operator performing this task. Although there is an established work cycle, it is often altered due to external influences, such as changes in geology, interruptions by coworkers and supervisors, machine malfunctions, and supply problems. The roof bolter operator is under pressure to install bolts as required to keep up with the coal mining operations while remaining vigilant to all possible dangers.

Figure 1.—Typical roof bolter posture.

Bolter operators must perform their tasks in an environment (Figure 2) confined vertically by the mine's roof height and horizontally by the bolter machine and closeness to the rib. Although Figure 2 shows an older roof bolter controls configuration, the environment is the same on newer equipment. This environment typically has low visibility due to the protective canopy and low light. This restricted work environment forces the operator into awkward postures for tasks that require quick reactions to avoid contact with moving machine parts. The bolting task requires working near unsupported roof, which increases the risk of injury from falling debris. Other factors affecting the operator's movement include wet or muddy conditions, uneven floor, and the required mining gear. In addition, the operator needs to maintain an arm's reach distance of 20–30 in from the moving boom arm because of the need to reach the controls and handle drill steels and bolts near the drill head. This requirement and the work environment force the operator to remain close to the moving boom arm.

Studies by MSHA [1994] and Turin et al. [1995] revealed there were no data for determining safe operating velocities for bolter arms operating close to workers in confined environments such as an underground coal mine. Ambrose et al. [2005] examined vertical boom velocity and provided valuable data and guidance. The results of this research provide the same information for horizontal roof bolter boom swing velocities.

Klishis et al. [1993b] examined worker job performance, risks, and hazard exposures during bolting operations. More than 12 bolting-related problems were identified as situations leading to injury. This study gave suggestions on how to avoid these situations which were subsequently evaluated at mining operations. Turin et al. [1995] conducted an analysis of hazards related to the movement of the drill head boom of a roof bolting machine. They recommended the following short-term solutions to increase the safety of roof bolter operators:

(1) Use an interlock device to cut off power to the controls when the operator is out of position.
(2) Place fixed barriers at pinch points and other dangerous areas.
(3) Provide better control guarding.
(4) Reduce the fast-feed speed.
(5) Use automatic cutoff switches for pinch points and other dangerous areas.
(6) Redesign the control bank to enable the operator to select the desired control by feel.
(7) Use resin insertion tools.

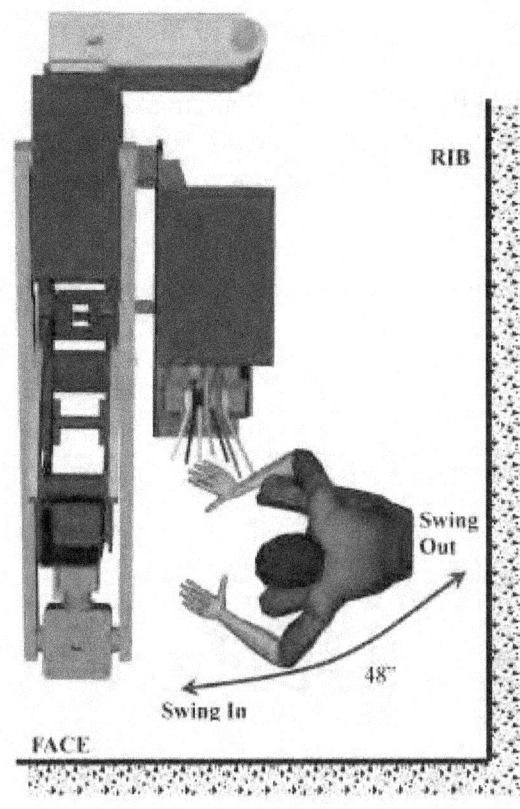

Figure 2.—Overhead view of bolting environment.

PHASE 1 TESTS: NIOSH VOLUNTEER SUBJECTS

Study Population

Since no special skills were involved for phase 1, eight volunteers from the NIOSH Office of Mine Safety and Health Research in Pittsburgh, PA, were used for the testing. The group consisted of seven males and one female ranging from 22 to 57 years of age and from 3rd to 99th percentile [McDowell 2005] by stature for males. The desired study population was the 5th through 95th percentile for males, representing 99% of the male target population. The objective of the laboratory tests was to ensure that the experimental design represented an accurate picture of the real world, not to duplicate the entire target population. Therefore, the small sample size including a female did not affect the goal of the tests.

The volunteers were solicited through a site-wide e-mail. Exclusion criteria for the testing included cardiovascular illness, knee problems, recent surgery or ailments, and current medications, which could limit or adversely affect the required movements.

Experimental Design and Measurements

The testing required that the subjects move along a 40-in-long arc path on the floor that simulated the movement of an operator during the boom swing function on a Fletcher Roof Ranger II bolting machine. Although a wooden mockup roof bolter of this make/model machine was available, it was not used because it was desired to have the subjects unencumbered by significant machinery in their environment. Instead, a test fixture (Figure 3) was designed and constructed of steel for the testing. It was a partial arc supported at both ends with a sliding lever and knob. The lever and knob moved an effective length of 50 inches in a pattern conforming to the motion of a roof bolter boom while swinging laterally.

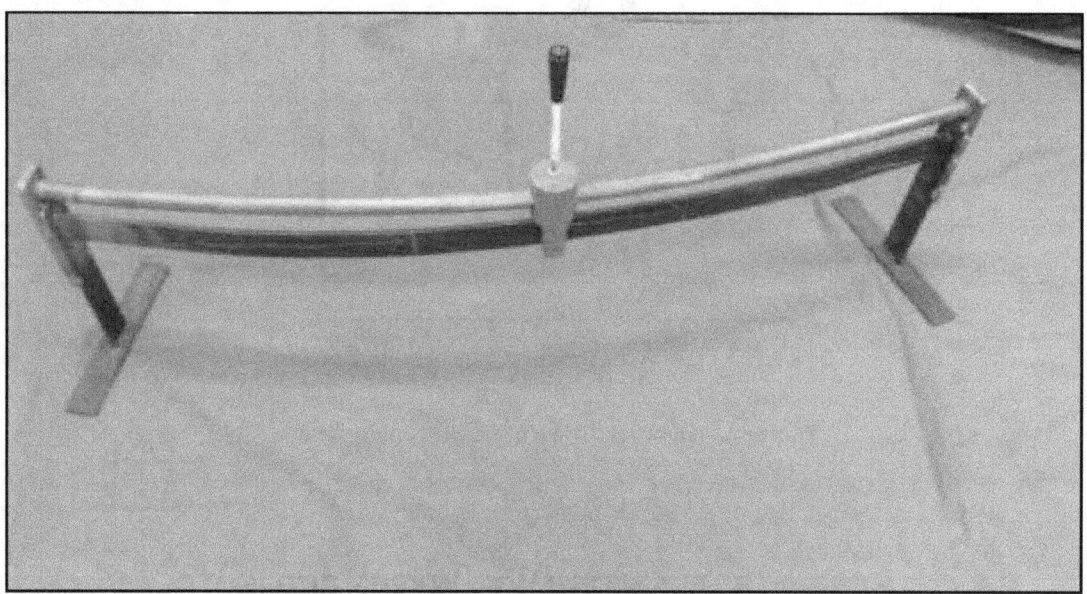

Figure 3.—Slide mechanism used for NIOSH subject testing.

The testing was intended to address these questions:

(1) What is the maximum velocity an individual perceiving danger can move toward or away from the bolter arm along a specified path in various work postures and seam heights?
(2) What effects do postures and seam heights have on an individual's motion and velocity?

The experiments were conducted using working heights of 72, 60, and 48 in. For the 72-in working height, a standing posture was used. For the 60-in height, stooping and squatting postures were used. For the 48-in height, squatting and kneeling postures were used. These four postures (standing, stooping, squatting, kneeling) are depicted in Figures 4–7. The roof height was adjusted by moving a suspended panel covered with a mesh fabric that represented the mine roof. Based on in-mine observations, the posture and working height combinations were thought to represent realistically what is experienced in real-world roof bolting situations.

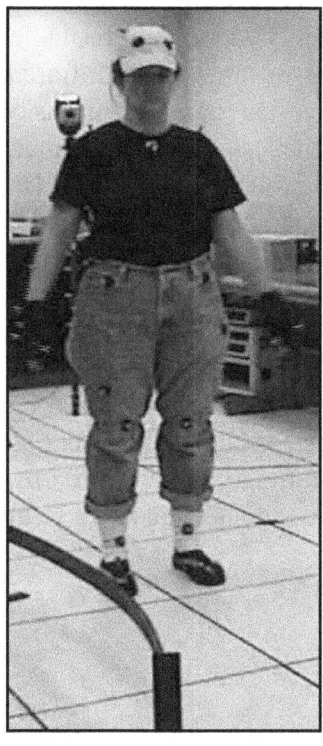

Figure 4.—Standing posture.

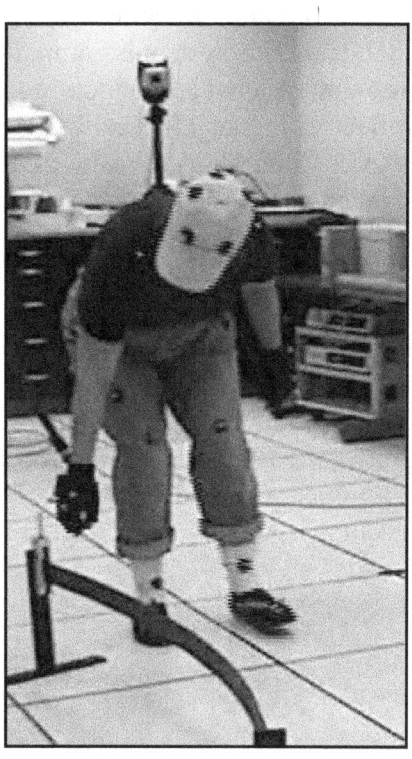

Figure 5.—Stooping posture.

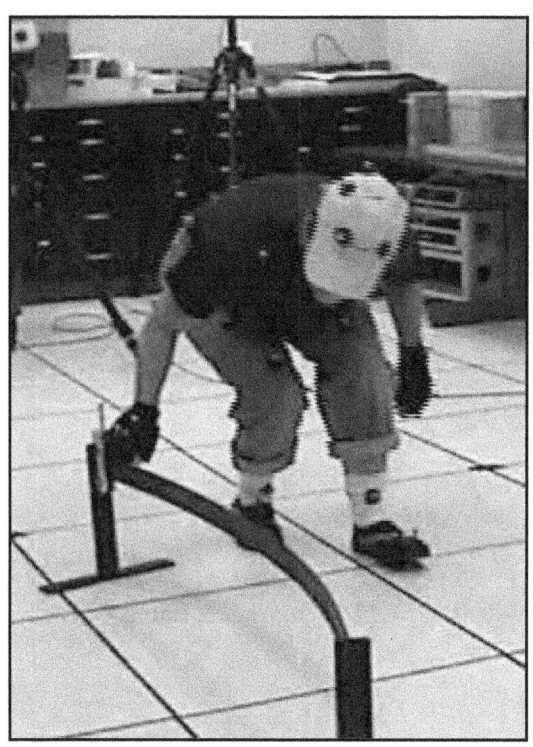

Figure 6.—Squatting posture.

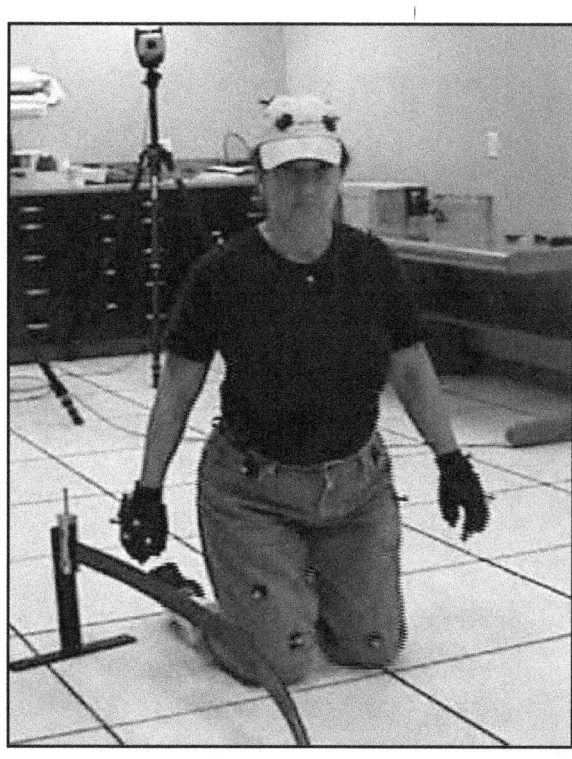

Figure 7.—Kneeling posture.

An optical motion tracking system by Motion Analysis Corp. was used to record the experimental data. This system is computer-based and is a highly accurate, repeatable, non-contact three-dimensional (3-D) measuring system that uses two-dimensional infrared camera views of reflective markers in a calibrated volume to calculate the relative positions of a suite of markers in three dimensions. This motion tracking system can be used to track humans, parts of equipment and machinery, and precision movements for human-task applications. The motion capture system uses an array of reflective markers that are placed on the subject and other items of interest. Although there are recommended arrangements of markers for human motion capture, this software is very flexible and allows use of any practical arrangement of markers. The array of markers used in this testing consisted of 41 markers called the JACK marker set. The locations and naming conventions for the 41 markers are shown in Figures 8–9. Using this marker set enabled compatibility with both the motion analysis software and a virtual environment program called JACK. This software analyzes human performance via modeling and virtual environment simulation. The marker set allows importation of motion capture data from test subjects into JACK and will enable future research to be conducted easily in virtual environments.

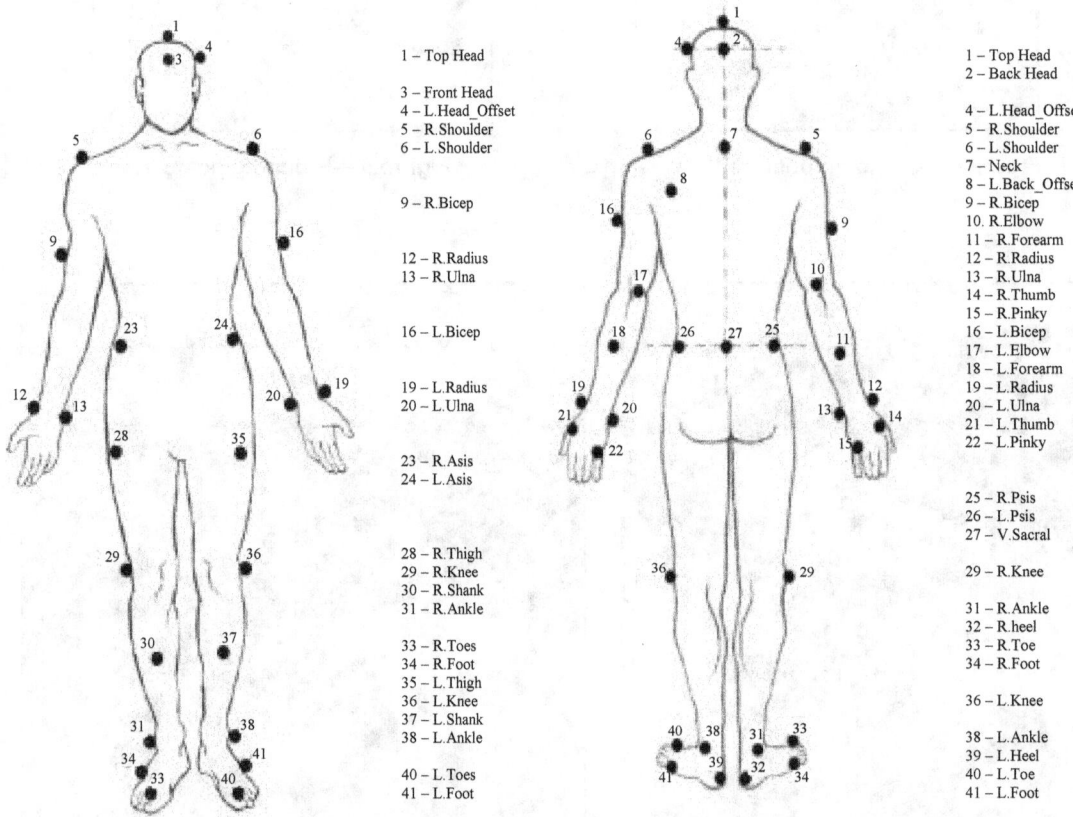

Figure 8.—JACK markers, front.   Figure 9.—JACK markers, back.

At the start of testing, the subjects were instructed to position themselves next to the test fixture to simulate working with a Fletcher Roof Ranger II machine. For the selected posture and working height, the subject was asked to assume as natural a position as possible with the right hand grasping the knob of the test fixture (Figure 10). At the verbal instruction—begin," the test subjects moved themselves forward along an arced path that simulated the motion of the bolter boom.

The subjects were told to move at a quick but comfortable pace while keeping their arm/hand and control lever/knob in the same position relative to their bodies. At the end of the forward excursion of the test fixture (about 50 in), the subjects were asked to remain stationary for 2 sec until they were given the verbal command—back." Then they repeated the exercise while moving backward. This procedure allowed the capture of both the forward and backward movements to be done in one recording session.

The sequencing of experimental conditions was randomized for each subject. Within each experimental condition, three repetitions of the tests were performed to obtain average values. Subjects were allotted 2 min of rest if desired between repetitions. These experiments were videotaped and photographed to ensure that the recorded data could be correlated with the events that occurred.

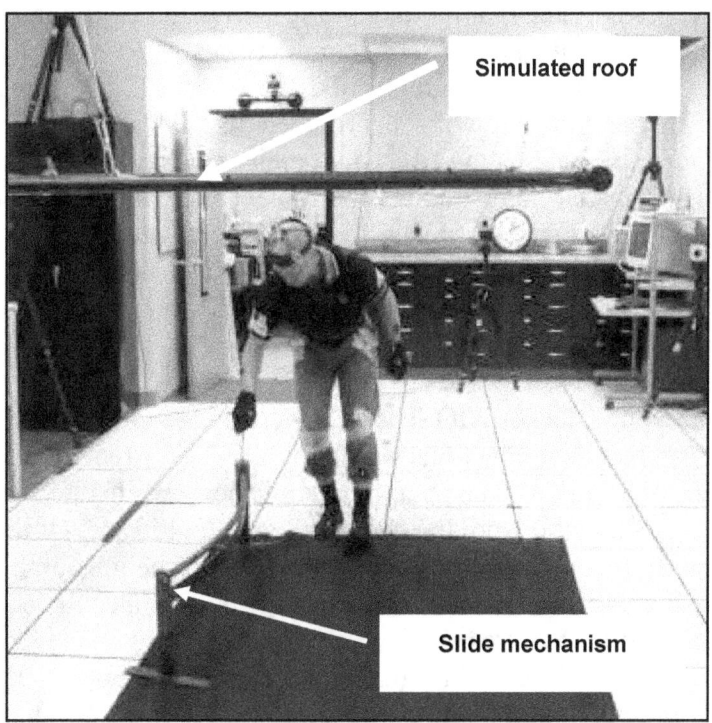

Figure 10.—NIOSH subject test fixture and artificial roof.

DATA ANALYSIS: NIOSH SUBJECTS

Methods

An analysis of the motion data determined the velocity at which subjects could move along the specified path for given postures and working heights. This information became the basis to set the boom swing velocity for the phase 2 tests with mine worker subjects. The motion data were also analyzed to try to answer the design study questions and to extract other items of interest.

As supplied by the manufacturer, a Fletcher Roof Ranger II bolting machine can swing its boom laterally from one extreme to the other (about 48 in) in 4 sec. This applies to the boom swinging both toward and away from the operator and is equal to a velocity of 12 in/sec. It was assumed that the subjects would be capable of moving along the bolter arm path at 12 in/sec for all of the posture and working height combinations. A primary goal of this testing was to determine an upper range of velocities, based on postures, working heights, and direction of movement, to be used in the second phase of experiments using the wooden mockup roof bolter and mine worker subjects.

The computer program Motion Analysis EVa Real-Time was used to determine exactly when the subjects started and ended their forward and backward motions. This program is part of the software library supplied by the motion tracking system manufacturer and was used to record the test sessions. The program allows the user to simultaneously view a 3-D depiction of all of the markers; a video of the subject; and a graphical display of the x, y, and z coordinates of a specific marker. Using this combination of synchronized displays and being able to step through the captured motion frame by frame allowed a very precise determination of when significant events occurred, such as on what frame number the subject started to move and on what frame number the subject's movement stopped. Since the frames were recorded at a constant rate of 60 frames/sec, it was simple to calculate the velocities for each trial and subject.

Results

Detailed results of the phase 1 NIOSH subject testing are presented in Table 1. The mean velocity for all postures/height combinations was 24.5 in/sec, over twice that of the base velocity for the bolting machine of 12 in/sec. The overall mean velocity while moving forward was 25.6 in/sec; the mean velocity while moving backward was 23.4 in/sec. Of the 120 tests conducted, there were only four instances of individual posture and height combinations where the subjects were not able to maintain the base velocity of the bolting machine at 12 in/sec. This occurred for the 72-in standing posture moving forward at 11.3 in/sec, the 72-in standing posture moving backward at 11.4 in/sec, the 48-in squatting posture moving backward at 9.4 in/sec, and the 48-in kneeling posture moving backward at 11.8 in/sec. Three out of four of these low velocities were for the 3rd-percentile subject, while the remaining one occurred with the 87th-percentile subject. Figures 11–12 are bar graphs depicting the statistical information contained in Table 1.

Table 1.—Detailed results of NIOSH subject testing

Percentile by stature	Subject	Trial	Forward postures: velocity, in/sec						Backward postures: velocity, in/sec					
			72 in, standing	60 in, stooping	60 in, squatting	48 in, squatting	48 in, kneeling	Mean	72 in, standing	60 in, stooping	60 in, squatting	48 in, squatting	48 in, kneeling	Mean
3	A	1	11.3	17.9	20.7	18.1	14.9		14.5	18.9	19.3	13.6	11.8	
		2	13.8	19.9	20.4	17.0	16.4		13.7	20.4	18.3	9.4	11.8	
		3	17.3	20.1	20.1	18.7	19.3		15.2	22.7	23.6	16.5	12.3	
		Average	14.2	19.3	20.4	18.0	16.9	17.7	14.5	20.7	20.4	13.2	11.9	16.1
87	B	1	14.8	16.0	18.6	21.3	17.6		13.3	14.7	16.2	19.2	20.3	
		2	14.5	18.6	20.7	19.2	20.1		11.4	14.8	18.8	18.0	16.1	
		3	14.6	16.4	20.3	19.1	17.1		13.2	17.0	16.6	16.6	17.5	
		Average	14.6	17.0	19.9	19.9	18.3	17.9	12.6	15.5	17.2	17.9	17.9	16.2
75	C	1	35.6	30.6	25.7	22.2	15.2		33.9	29.7	24.8	24.2	18.3	
		2	35.6	32.7	28.2	27.4	21.8		28.2	29.4	29.1	27.4	19.9	
		3	42.4	25.9	33.9	26.4	20.1		35.1	28.5	26.2	30.6	13.5	
		Average	37.8	29.8	29.3	25.3	19.0	28.2	32.4	29.2	26.7	27.4	17.2	26.6
62	D	1	20.6	24.4	24.6	26.9	17.5		18.5	26.2	22.2	22.9	12.5	
		2	26.7	30.6	24.4	22.5	19.5		24.2	28.5	26.7	21.0	15.9	
		3	26.4	25.9	21.3	26.7	17.3		26.7	25.9	26.2	15.8	17.1	
		Average	24.6	27.0	23.5	25.4	18.1	23.7	23.1	26.9	25.0	19.9	15.2	22.0
14	E	1	39.5	29.4	38.9	27.7	21.7		31.3	28.8	33.5	31.3	19.6	
		2	38.4	29.1	36.9	36.9	20.9		36.0	30.6	30.6	34.7	24.6	
		3	41.1	32.0	40.0	38.9	31.0		35.1	30.3	31.0	34.7	17.3	
		Average	39.7	30.2	38.6	34.5	24.5	33.5	34.1	29.9	31.7	33.6	20.5	30.0
9	F	1	21.3	25.3	28.5	18.9	25.3		19.5	26.2	24.4	23.6	13.6	
		2	29.7	22.2	28.0	25.0	23.6		25.7	24.2	24.2	24.6	14.8	
		3	29.1	26.9	26.7	25.0	18.1		28.0	25.7	26.9	25.3	14.3	
		Average	26.7	24.8	27.7	23.0	22.3	24.9	24.4	25.4	25.2	24.5	14.3	22.7
99	G	1	32.0	36.9	33.9	31.6	27.2		38.4	35.6	27.4	31.6	23.2	
		2	35.1	41.1	29.1	32.0	23.4		29.7	40.0	32.0	29.4	20.4	
		3	37.9	43.0	34.7	38.9	26.2		34.7	43.6	29.4	33.1	28.5	
		Average	35.0	40.4	32.6	34.2	25.6	33.5	34.3	39.7	29.6	31.4	24.1	31.8
71	H	1	23.2	21.3	25.5	23.8	16.9		17.0	22.2	24.4	22.0	16.8	
		2	26.4	28.8	28.8	25.0	26.9		19.6	22.2	25.3	26.4	22.2	
		3	22.2	23.6	29.1	28.2	25.3		18.7	25.0	23.8	22.2	15.7	
		Average	23.9	24.6	27.8	25.7	23.0	25.0	18.4	23.1	24.5	23.5	18.2	21.6
Statistics		Mean	27.1	26.6	27.5	25.7	21.0	25.6	24.2	26.3	25.0	23.9	17.4	23.4
		SD	9.7	7.3	6.3	6.3	4.3		8.7	7.0	4.7	6.9	4.3	
		Maximum	42.4	43.0	40.0	38.9	31.0		38.4	43.6	33.5	34.7	28.5	
		Minimum	11.3	16.0	18.6	17.0	14.9		11.4	14.7	16.2	9.4	11.8	

SD Standard deviation.

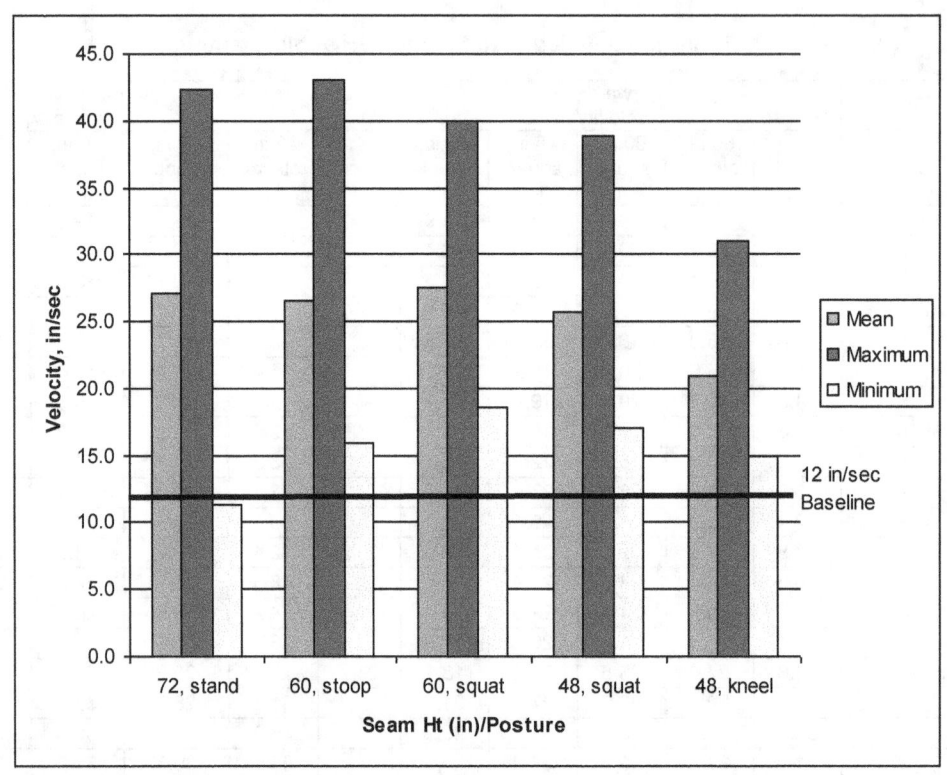

Figure 11.—Velocities for forward boom swing motions.

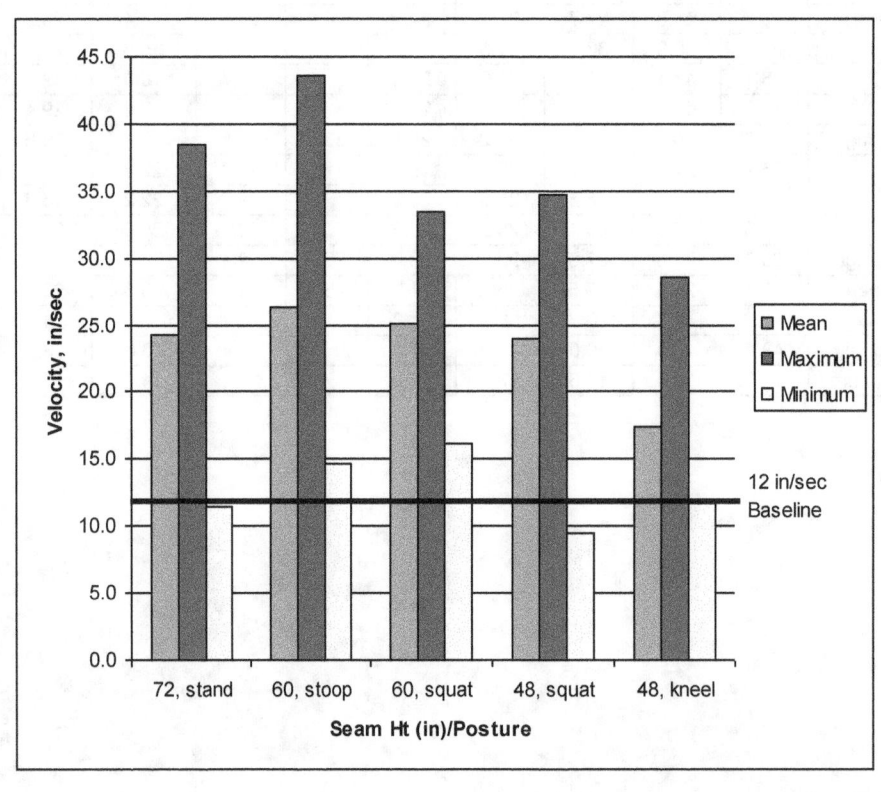

Figure 12.—Velocities for backward boom swing motions.

For all of the subjects, the highest mean velocity was demonstrated at the 60-in squatting posture while moving forward (27.5 in/sec). The posture that showed the lowest mean velocity was the 48-in kneeling posture while moving backward at 17.4 in/sec.

For the range of working heights and postures tested, there was a positive correlation between the subject's height and average velocity. Graphs were prepared plotting subject height versus the average recorded velocity for all of the tested working height and posture combinations. Figures 13–22 show examples of these plots for all of the postures, with the forward and backward motions shown separately. Also shown is a line that is fitted to the data using regression analysis, the equation for the line, and the correlation coefficient. For all of the cases, the lines showed positive slopes. The correlation coefficients ranged from 0.009 to 0.552.

Based on this testing, it was decided to conduct the phase 2 testing with the mockup roof bolting arm so that the velocity of the arm would be 12 in/sec, 16 in/sec, and 24 in/sec when moving away from the operator and 12 in/sec, 13.7 in/sec, and 16 in/sec when moving toward the operator. The manufacturer's base velocity of 12-in/sec was selected as the slowest velocity for both directions of boom arm testing. As detailed above, of the 120 tests conducted, there were only four instances where the tested subjects were not able to maintain the manufacturer's standard velocity of 12 in/sec. Therefore, test velocities below 12 in/sec were not deemed to be necessary. Although maximum velocities of over 30 in/sec were recorded for all tested postures for the swing-in (operator walking forward) direction, this velocity was tested with the mockup boom arm and it was determined that 30 in/sec would be excessive and hazardous to test subjects. Therefore, 24 in/sec was selected as a fast swing-out velocity as it proved to be a challenging velocity as supported by the previous analysis. The velocity of 16 in/sec was selected as a fast velocity for the backward motion because this was near the mean value calculated for the kneeling posture at 17.4 in/sec. The decision was made to limit the number of velocities to three, since changing the velocities on the test apparatus was a cumbersome and somewhat imprecise procedure that used a stopwatch to time the swing from start to stop. We chose the medium velocity to be halfway between the fast and normal velocities—13.7 in/sec or 3.5 sec for the swing-out and 16 in/sec or 3.0 sec for the swing-in.

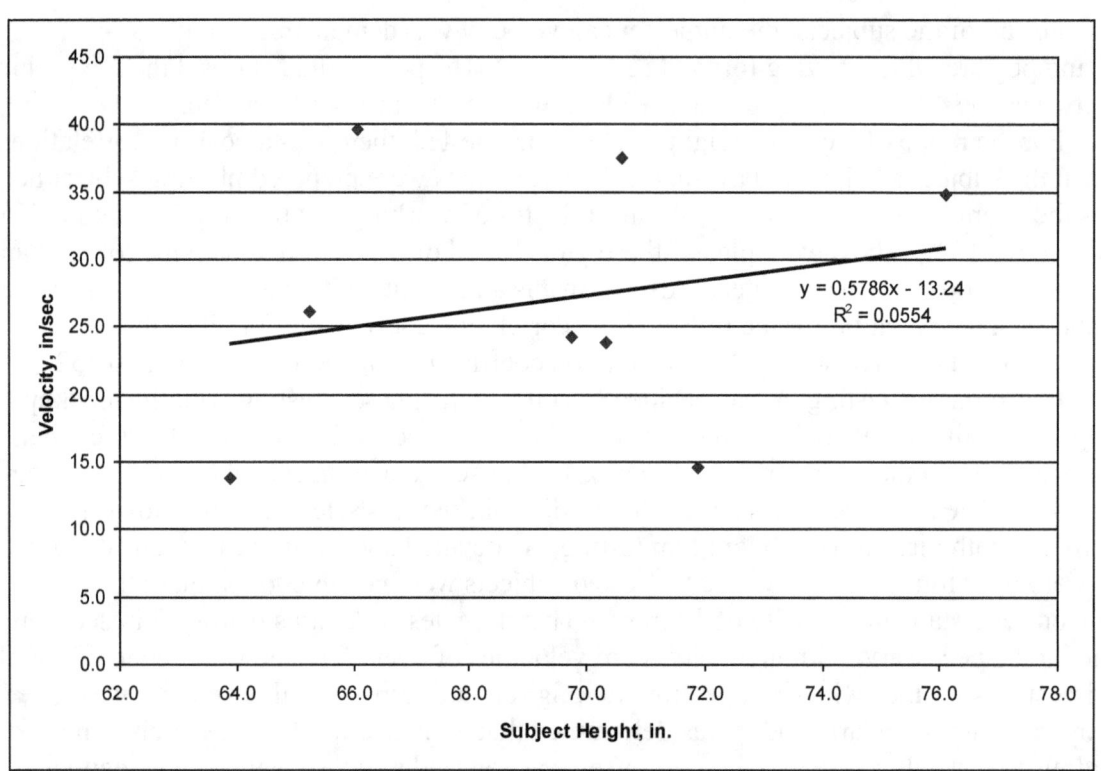

Figure 13.—NIOSH subjects: 72-in seam height, standing posture, forward motion.

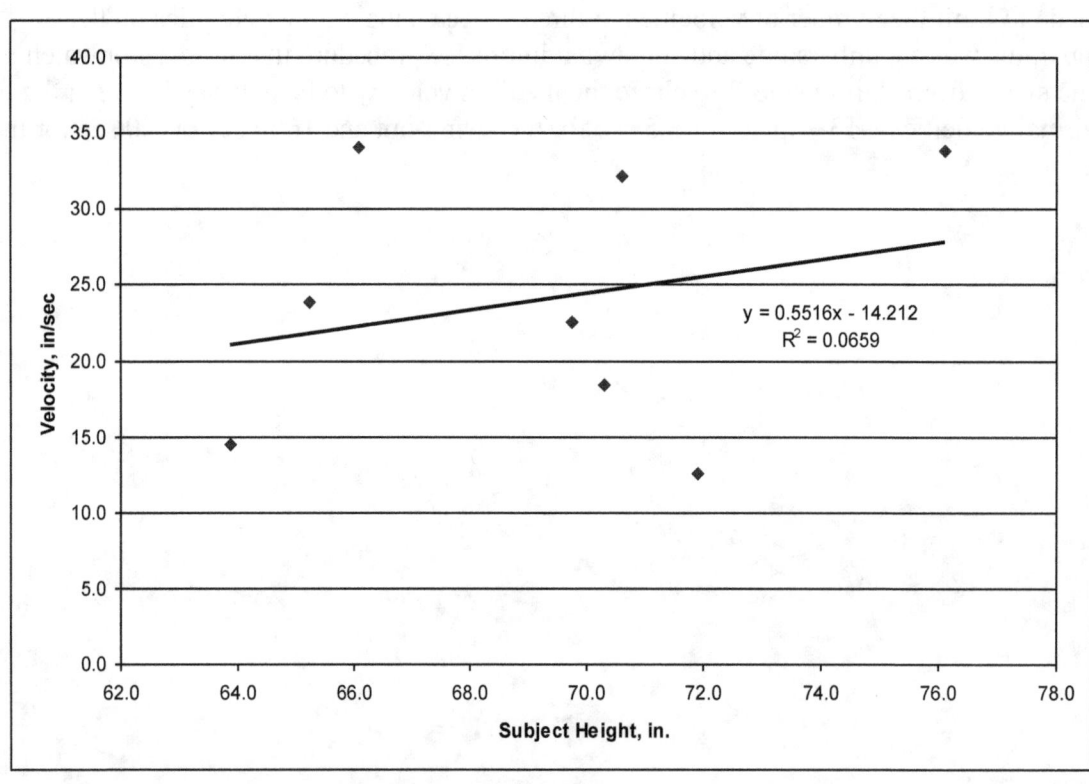

Figure 14.—NIOSH subjects: 72-in seam height, standing posture, backward motion.

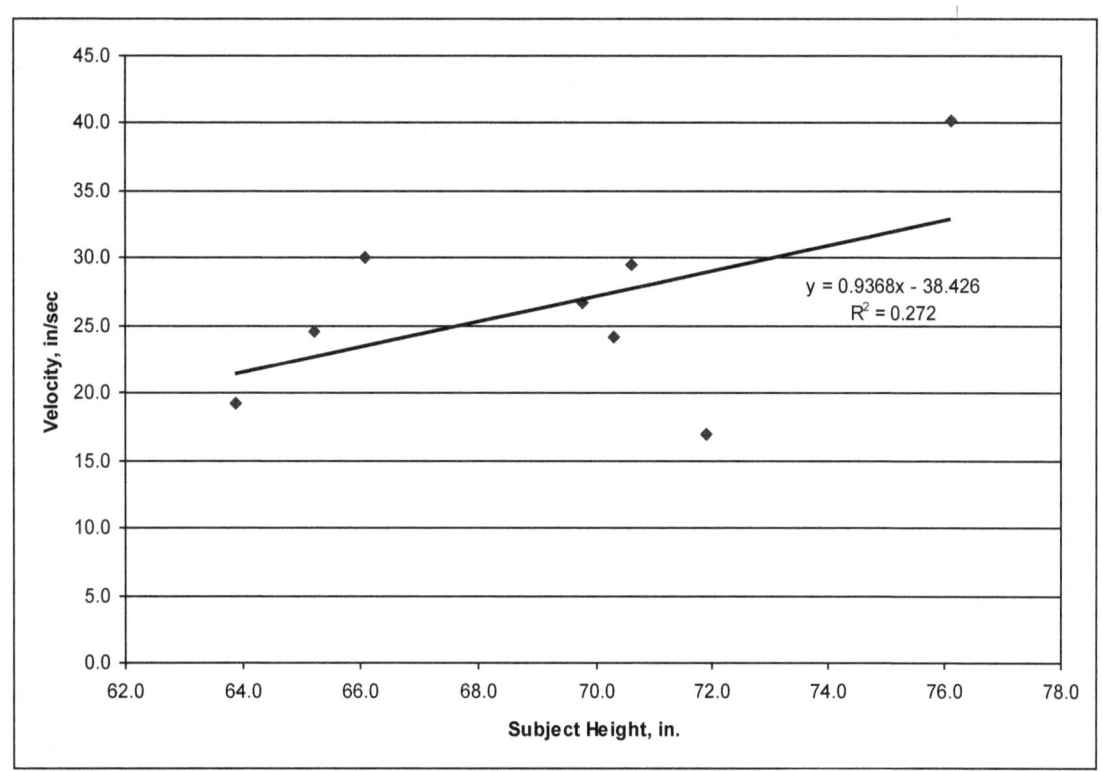

Figure 15.—NIOSH subjects: 60-in seam height, stooping posture, forward motion.

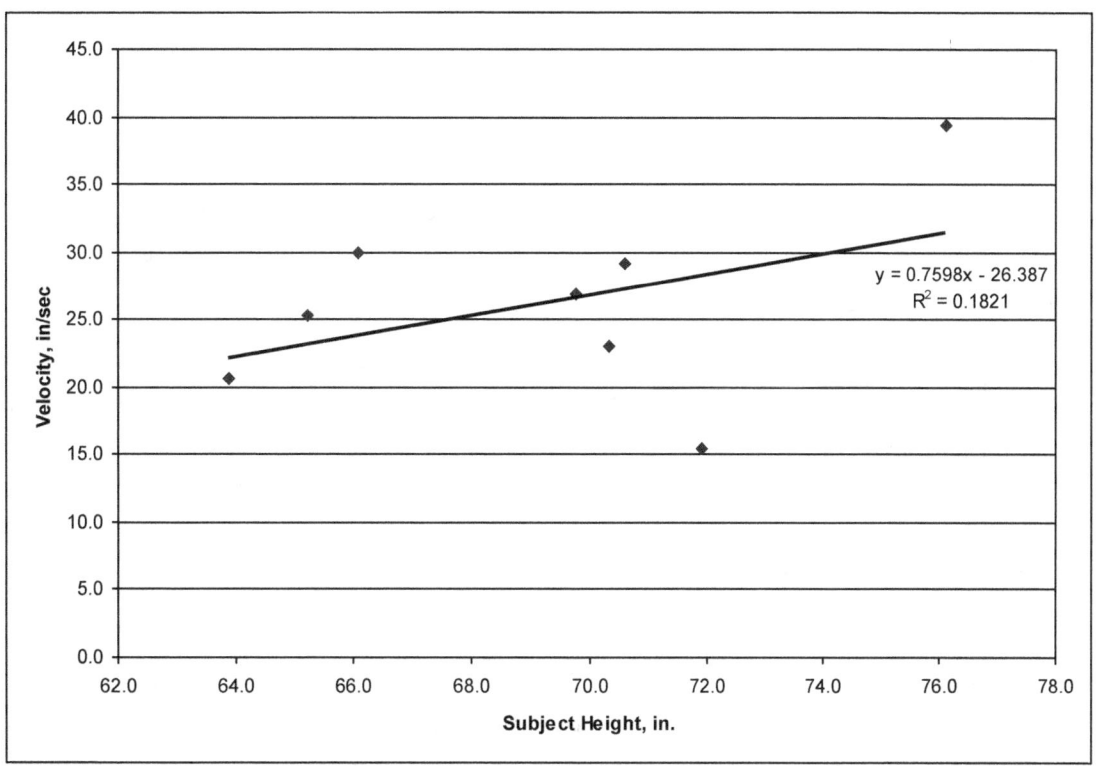

Figure 16.—NIOSH subjects: 60-in seam height, stooping posture, backward motion.

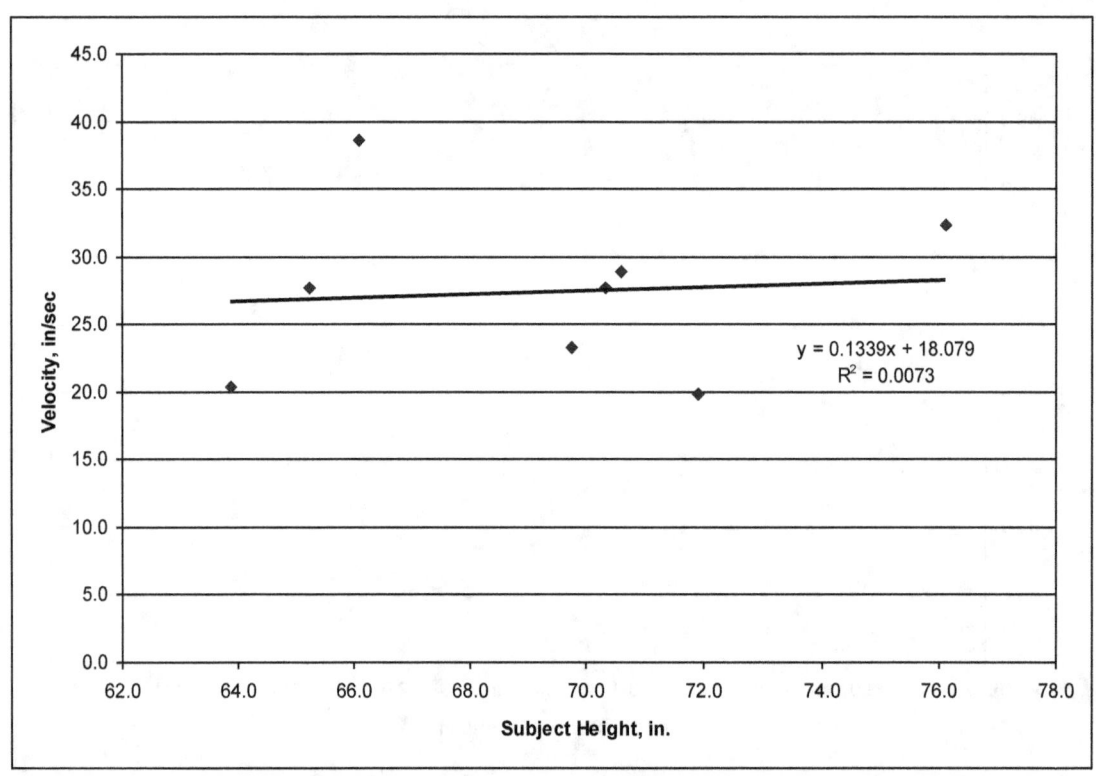

Figure 17.—NIOSH subjects: 60-in seam height, squatting posture, forward motion.

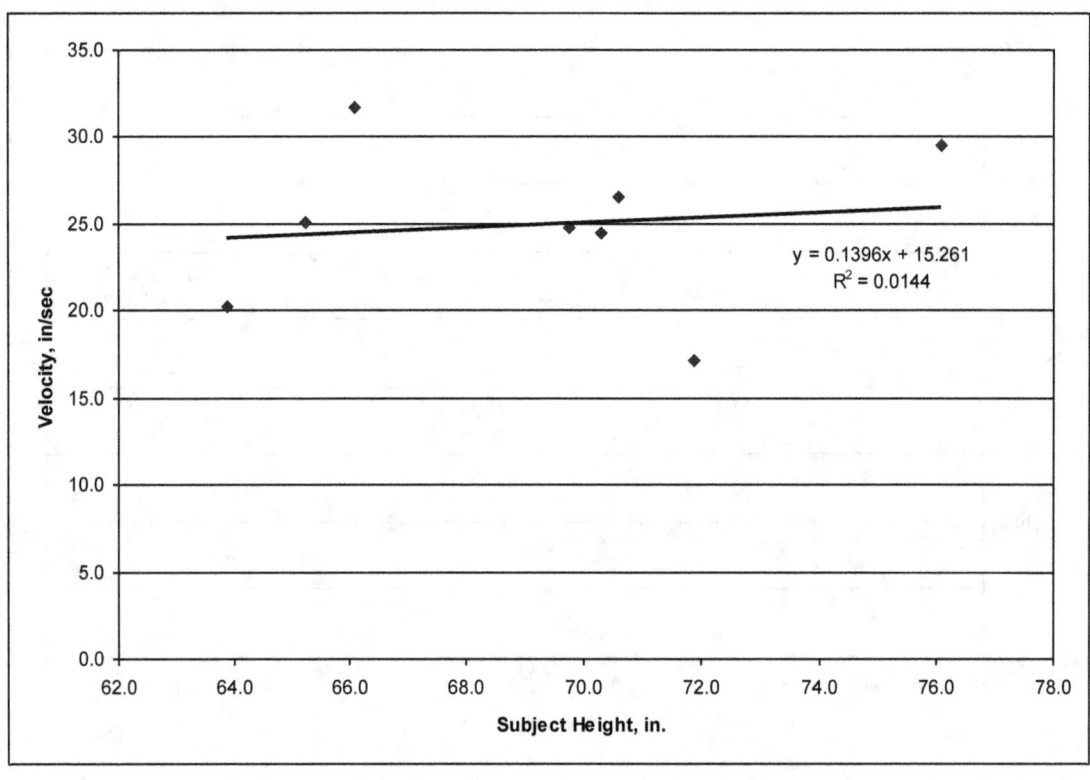

Figure 18.—NIOSH subjects: 60-in seam height, squatting posture, backward motion.

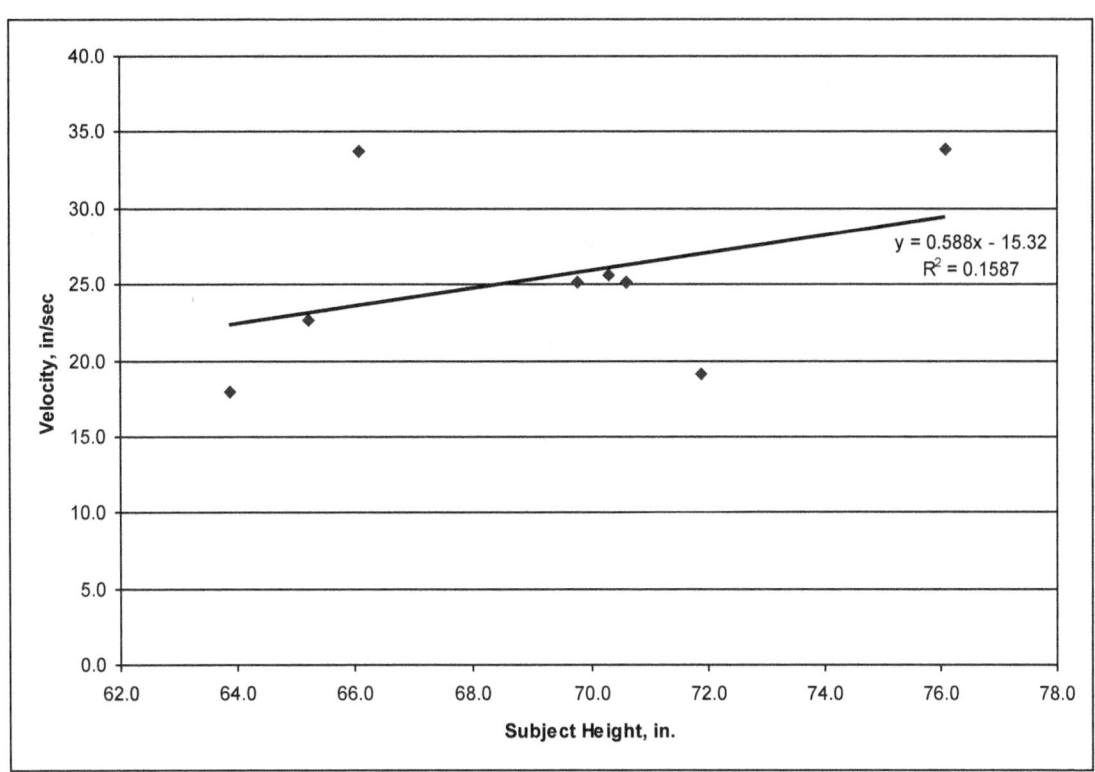

Figure 19.—NIOSH subjects: 48-in seam height, squatting posture, forward motion.

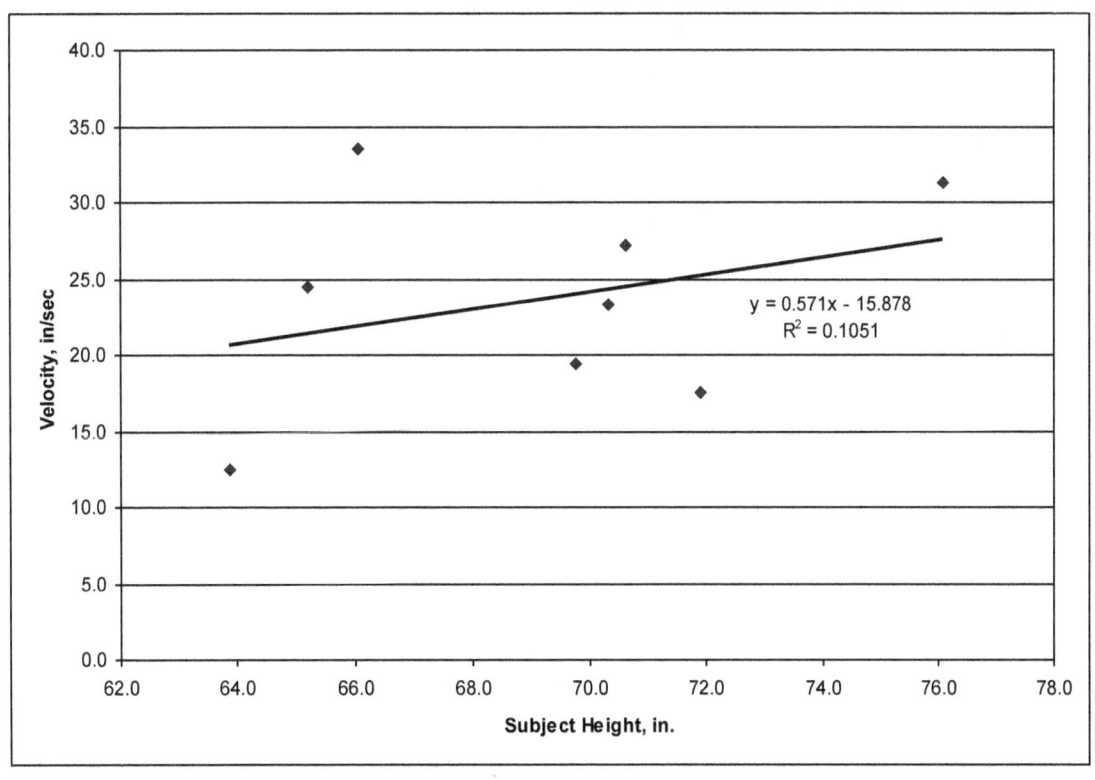

Figure 20.—NIOSH subjects: 48-in seam height, squatting posture, backward motion.

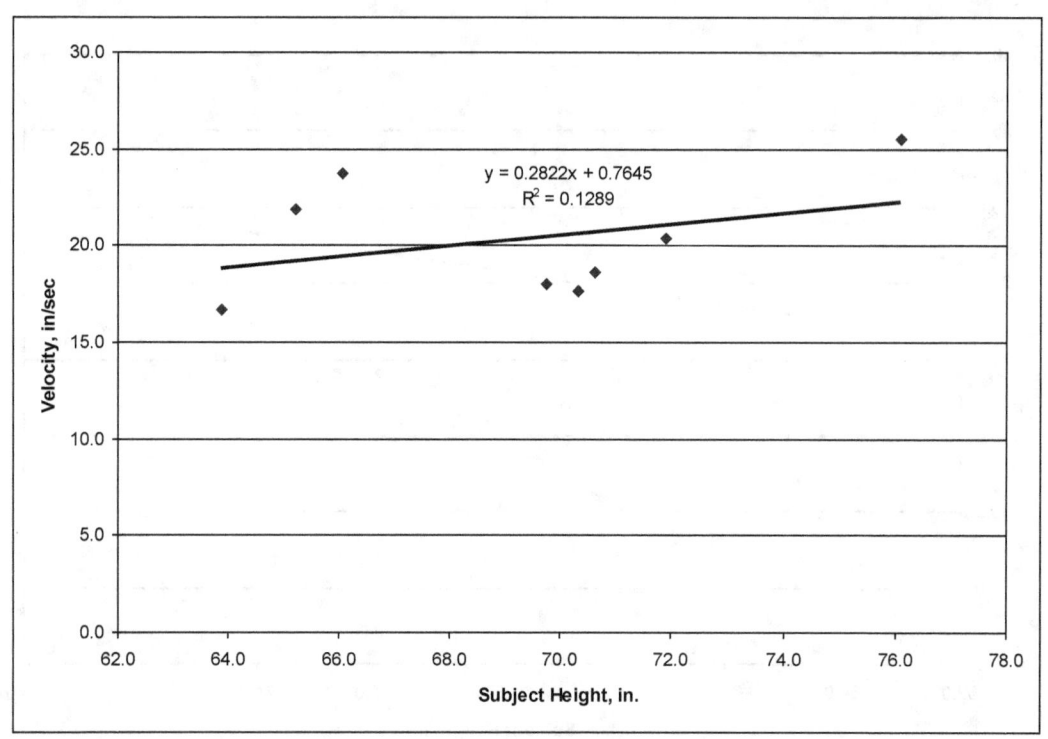

Figure 21.—NIOSH subjects: 48-in seam height, kneeling posture, forward motion.

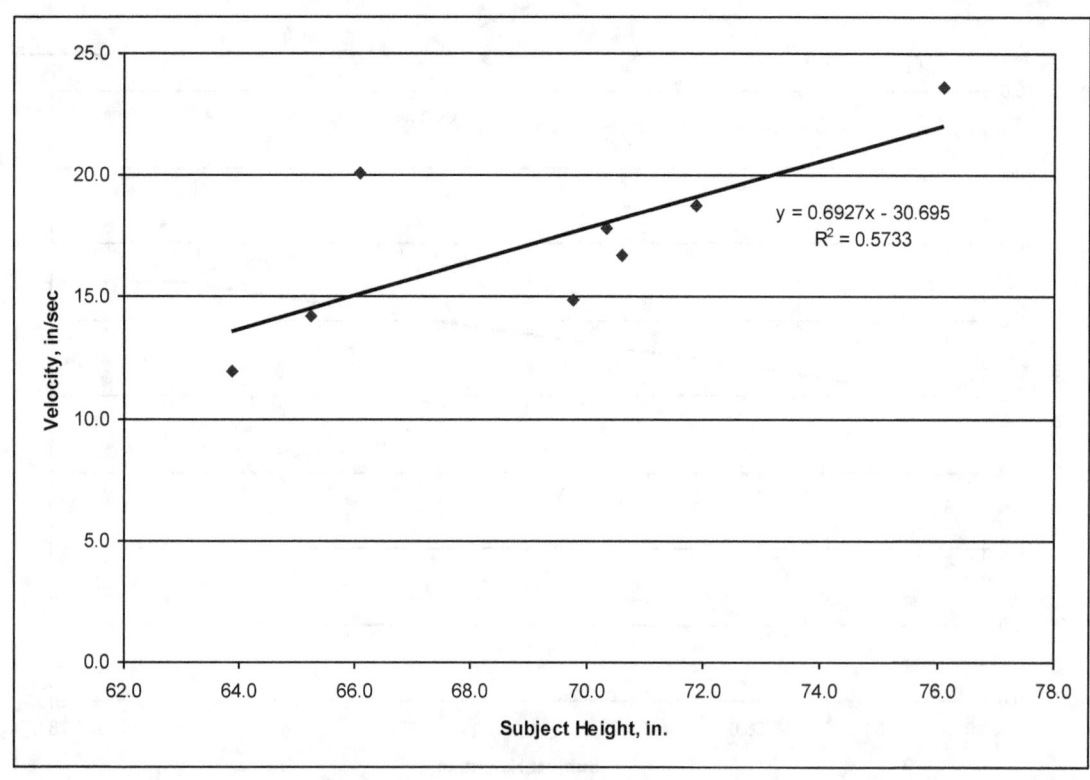

Figure 22.—NIOSH subjects: 48-in seam height, kneeling posture, backward motion.

Discussion

The statistics generated for the forward and backward motions did not differ significantly. As shown in Table 1, the average mean velocities for the forward motions were consistently higher than the corresponding backward motions regardless of posture. However, the overall forward motions were only 9.3% faster than the overall backward motions. The biggest difference was noted for the kneeling posture, where the forward motion was 20% faster than the backward motion. The smallest difference was noted for the 48-in squatting posture, where the forward motion was 7.6% faster than the backward motion.

The positive correlations between a subject's height and average velocity, as presented above and shown in Figures 13–22, can be explained by the fact that the larger subjects tended to have longer legs and could move through the distance of the test arc in one or two steps. The smaller subjects had shorter legs and strides that required three or more steps to complete the motions. For the small distance used in this testing, taking one or two steps was more efficient and faster than three or more steps.

Additional findings based on the subjects' anthropometry were initially desired but were found to be beyond the scope of this testing. A significant number of additional subjects would be required, and these subjects would need to exhibit a wide range of anthropometric differences.

PHASE 2 TESTS: MINE WORKER SUBJECTS

Study Population

Twelve mine worker volunteers participated in the study. Seven of the subjects were experienced bolter operators and two were mechanics. The remaining three listed their positions as general laborers. The volunteers were all males from 27 to 69 years of age and averaged about 46 years old. Their heights ranged from 5'5" to 6'3" (average 5'10"), and they weighed 160–270 lbs (average 210 lbs). Table 2 lists the subjects' anthropometric data. With one exception, the subjects were right-handed. There were no measurements of the subject's physical strength or motor skills because operating a roof bolting machine does not require more than nominal effort and coordination. Although this sample size is small, we believe it to be representative of the mining population. Interviews revealed both mechanics and two of the bolter operators knew that the roof bolter swing velocity could be adjusted by altering the setting of hydraulic flow control valves at any time. All subjects stated they had noticed variations in the rates at which different equipment of the same model operated.

Table 2.—Mine worker subject anthropometric data

Subject	Height, in	Weight, lb	Age, years	Operator percentile
1	72.7	190	55	93
2	68.4	236	55	36
3	69.3	230	52	52
4	68.2	200	57	35
5	68.5	240	38	38
6	68.7	160	57	43
7	70.0	205	54	66
8	72.4	185	49	91
9	72.2	175	27	89
10	75.4	270	52	99
11	67.6	245	31	27
12	65.3	178	69	10

Experimental Design and Measurements

The purpose of this study was to determine if the boom arm swing velocity, in combination with operator posture, is the most significant factor that affects the operator's ability to perceive and avoid machine contact hazards. Motion data were collected for human subjects using a full-scale working mockup of a roof bolter boom.

The test setup was a wooden mockup of a 72-in Fletcher Ranger II roof bolter left-hand arm. This make and model were chosen because of their overwhelming dominance in the industry. The arm was placed in full extension for all testing. To minimize risk to test subjects, various safety features were designed into the mockup. These included friction clutches on all actuators and laser proximity detectors. Figure 23 shows a subject at the start of a test in a stooping posture. Figure 24 shows a subject at the start of a test in a kneeling posture.

Table 3 lists the target times for a full swing during a test. The target velocities were based on the phase 1 tests and manufacturer data. The velocities were all set as a time base (i.e., seconds for full swing). The Fletcher Roof Ranger II bolting machine, as supplied by the manufacturer, can swing its boom laterally from one extreme to the other in 4 sec. With the 72-in boom length and the sump cylinder fully extended, the operator's station moves about 48 inches in an arc path during a full swing. This motion equates to a velocity of 12 in/sec and became the "normal" velocity used for the mine worker tests. Since the test had a fixed distance for the subject to move, time and velocity are transposable, as seen in Table 3.

Figure 23.—Mine worker subject in a stooping posture.

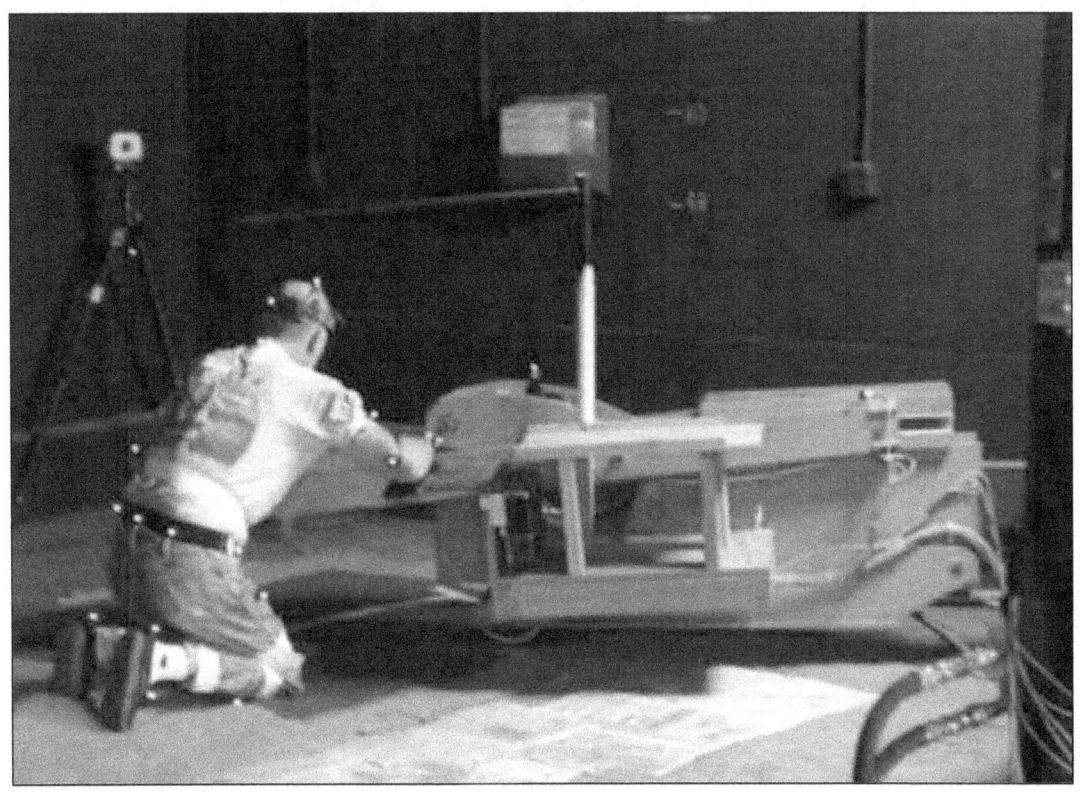

Figure 24.—Mine worker subject in a kneeling posture.

Table 3.—Target swing velocities and times

	Swing-in		Swing-out	
	Target	Actual (average)	Target	Actual (average)
Normal:				
Time, sec	4.0	3.7	4.0	3.8
Velocity, in/sec	12.0	12.8	12.0	12.6
Medium:				
Time, sec	3.0	2.7	3.5	3.2
Velocity, in/sec	16.0	17.6	13.7	14.9
Fast:				
Time, sec	2.0	2.2	3.0	2.7
Velocity, in/sec	24.0	21.5	16.0	17.5

The target velocities were set before motion capture sessions using a stopwatch while adjusting flow controls built into the bolter arm mockup's hydraulic system. At the time, it was thought this method provided sufficient accuracy. After testing with all mine worker subjects had been completed, analysis of the data showed that there was more-than-expected variance in the target swing velocities. So, rather than assuming the target velocities had been achieved, the data analysis had to take into account the velocity variances that had occurred from the desired target velocities. The actual times and velocities are also shown in Table 3.

The same optical tracking hardware and software used for the NIOSH subject tests were used for the mine worker tests. Several markers were placed on the boom arm to record its motion in addition to the standard 41-marker set on the subject. These trials were conducted at the same working heights and postures as those used in the NIOSH tests. There was no provision for a suspended ceiling. Also, the protective canopy normally covering the operator had to be removed to allow a clear line of sight for the optical tracking system's cameras. Instead, a small PVC tube was attached to the boom arm and thus moved with it. The tubing was used to simulate the obstruction created by the actual roof bolting machine's protective canopy. The height of this was adjusted appropriately for each test. As with the NIOSH tests, the sequence of velocity, height, and posture combinations was randomized. The motion capture system recorded video as well as motion data.

The testing consisted of having the subject perform the swing function under various working height and posture combinations. The following simulated working heights and associated postures were tested: 72-in standing, 60-in stooping, 60-in squatting, 48-in squatting, and 48-in kneeling.

DATA ANALYSIS: MINE WORKER SUBJECTS

Methods

Analysis of the motion capture data determined a rate of change of the distance between the operator and the moving boom arm through its 48-in swing. This analysis developed a visual picture of positional changes between the subject and the boom arm during the boom swing motion. A custom computer program was written in Visual Basic .NET to perform the initial analysis of the data. For each session, the motion capture system created an ASCII delimited file that contained the 3-D positions of each marker over the course of the observation. The program first identified the start and stop points for the boom while swinging out toward the operator and while swinging in, away from the operator.

Statistical information was calculated for each trial that included the average distance between two markers during a swing-in or swing-out operation, the standard deviation, the minimum, the maximum, and the range. The markers used for this were the boom arm marker closest to operator and the marker on the left hip of the subject, identified as the left ASIS marker (L.Asis - marker 24) in Figure 8. The program also calculated a linear regression for the distances versus time, with a record made of the resulting slope and the coefficient of correlation. The program calculated and compiled this information for all of the 45 data files that comprised the complete motion capture session for an individual subject. This information was placed in another delimited text file that was then imported into an Excel spreadsheet for further analysis.

The following is a simplified explanation of the statistical methods. Plotting the data in Excel showed a graphical representation of the distance between the two markers versus time. If, on average, this distance increased during the range of the boom swing, it meant that the boom arm tended to move away from the subject. If it decreased, it meant that the boom arm moved closer to the subject. Of prime importance was calculating the rate of change of distance between the operator and the boom arm. This rate of change is analogous to the velocity of the boom arm approaching or moving away from the moving operator. This is easily expressed as a slope, where the slope would be zero if the subject tended to keep an even pace with the moving boom arm.

By plotting this distance calculation versus time, graphs were produced using Excel, such as those shown in Figures 25–26. For these graphs, the slope of a line (the X coefficient) fitted to the data was associated with the tendency of the subject to move toward or away from the boom arm during a swing motion. A positive slope indicated that the subject and boom arm tended to separate. A negative slope indicated that the distance between the subject and boom arm closed. As a slope became more positive or negative, the rate of separation or closure increased. Although Excel was used to calculate the line slope in this explanation, the actual slopes and regressions used in the analysis were calculated by the custom computer program.

During this analysis, it was realized that the time intervals from the selected start and stop points of the swing motions varied between operators. This meant that the desired target velocities had not been exactly set during testing. To take this into account, the actual velocities were extracted from the data files, and the slope data were plotted against the extracted velocities instead of using target velocities for the x-axis of the final result graphs.

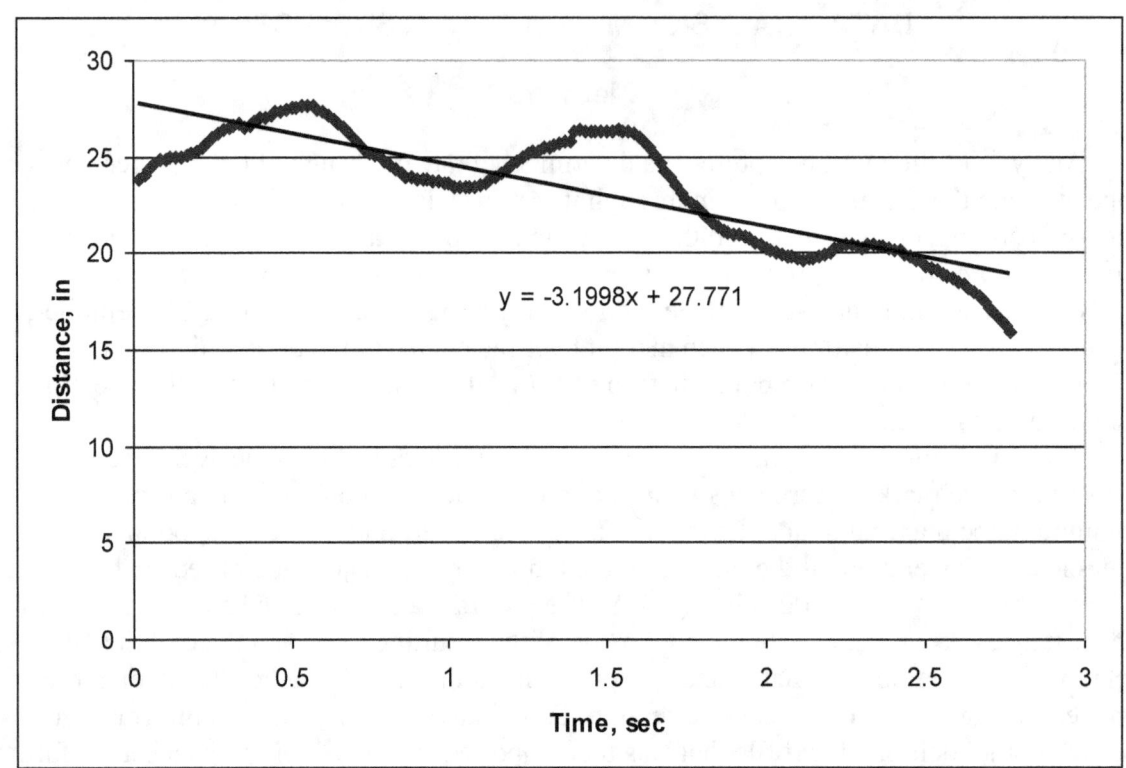

Figure 25.—Distance versus time plot for swing-out.

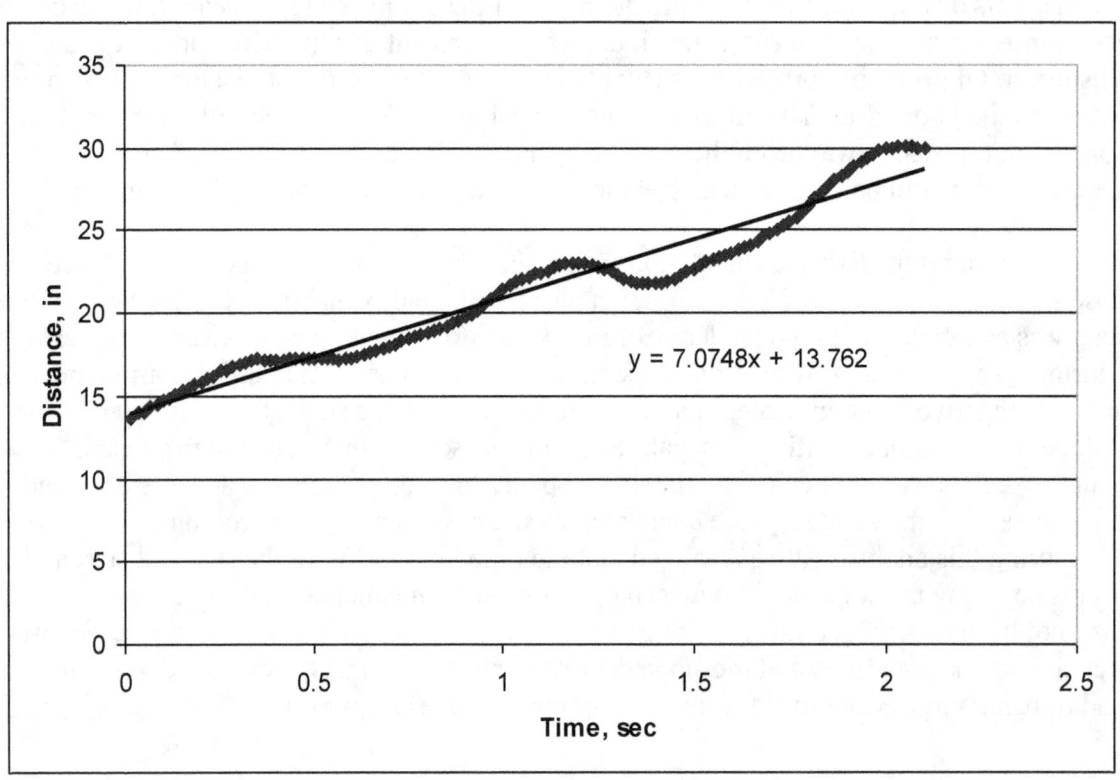

Figure 26.—Distance versus time plot for swing-in.

Results

The slope and swing velocity values from the custom software were imported into Excel for plotting and further analysis. The data are shown in the Appendix. Those results are depicted as X-Y scatter charts for each of the five working postures. To generate the final results, data for all 12 subjects were averaged together. Figures 27–31 show the results for each operating posture. The velocities are expressed as a slope to enable normalization of the data for individual operator physical characteristics. When the value of slope is positive, the subject-to-boom arm distance was increasing. Conversely, a negative slope indicates the subject-to-boom arm distance was decreasing or closing. As the absolute value of the slope increased from zero, the rate of change in distance between operator and boom arm increased.

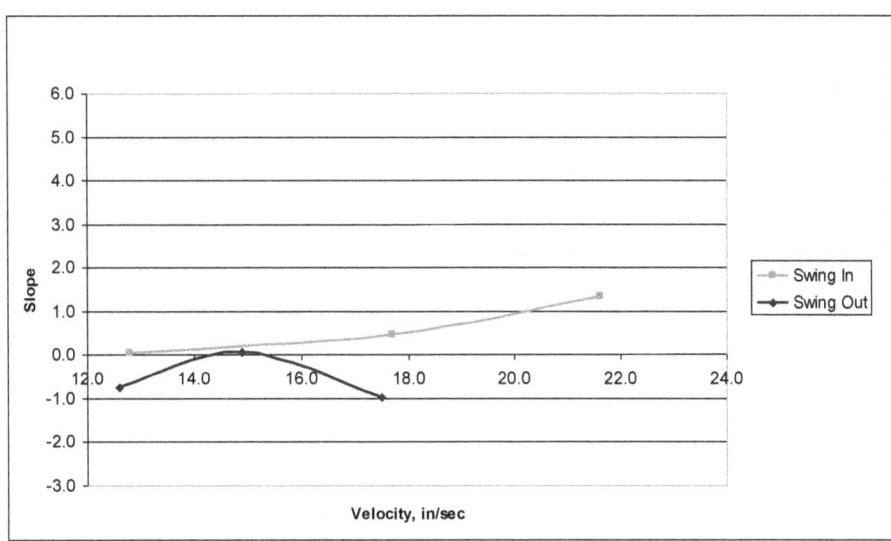

Figure 27.—Mine worker subjects: standing 72-in results.

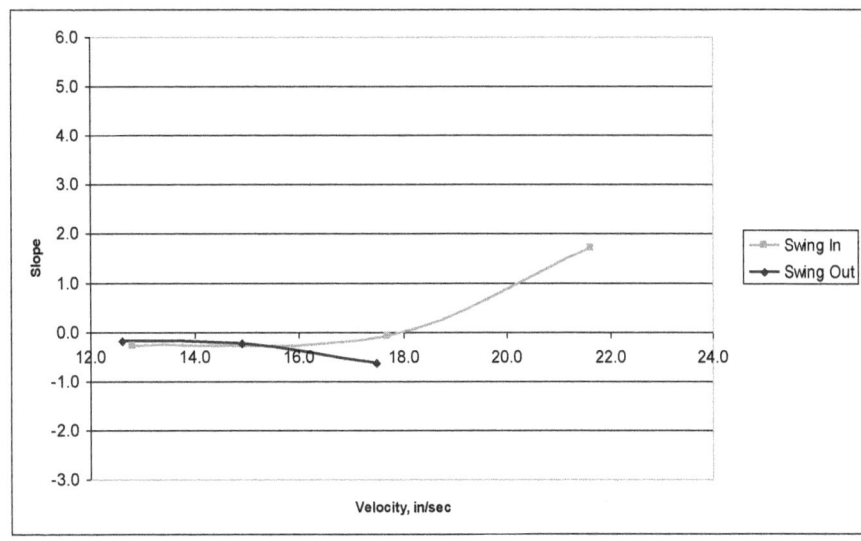

Figure 28.—Mine worker subjects: stooping 60-in results.

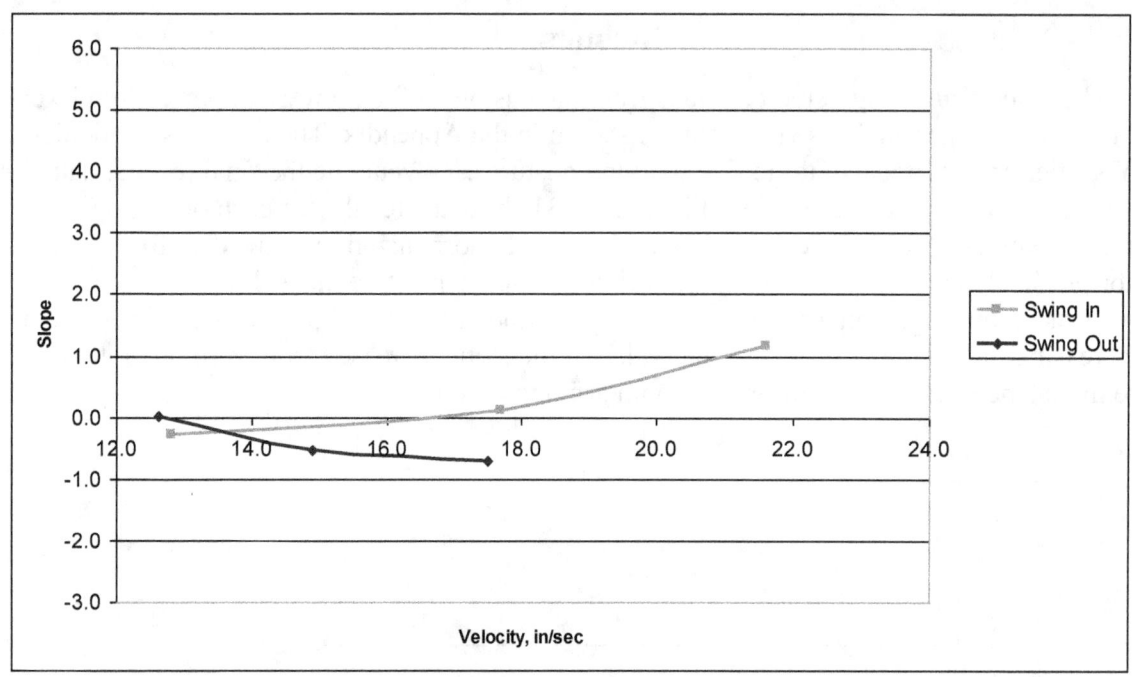

Figure 29.—Mine worker subjects: squatting 60-in results.

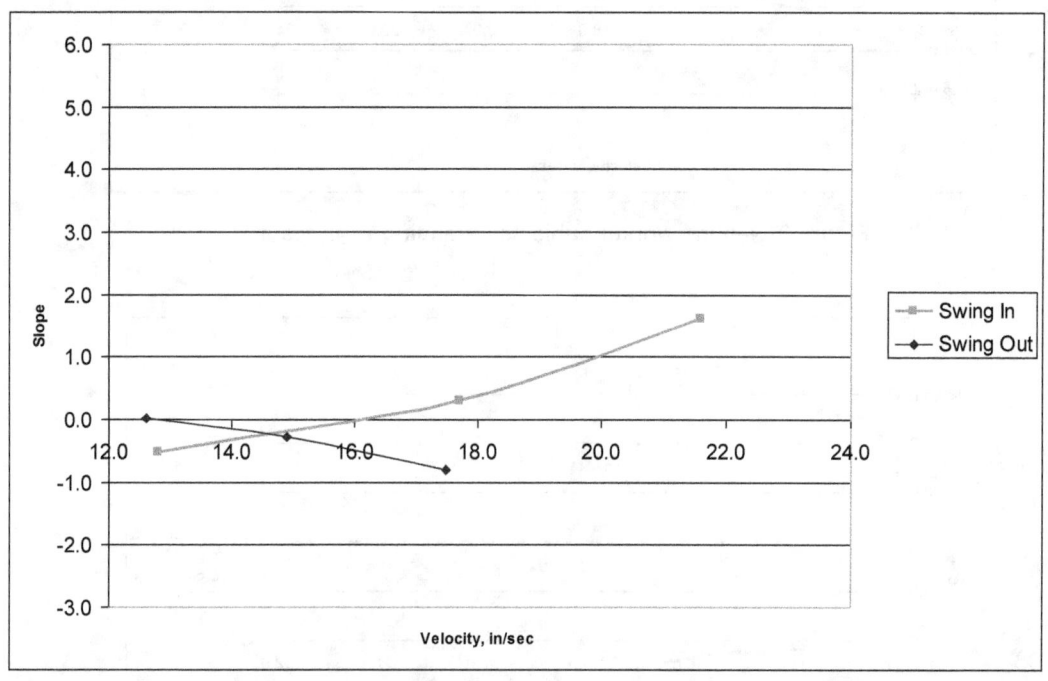

Figure 30.—Mine worker subjects: squatting 48-in results.

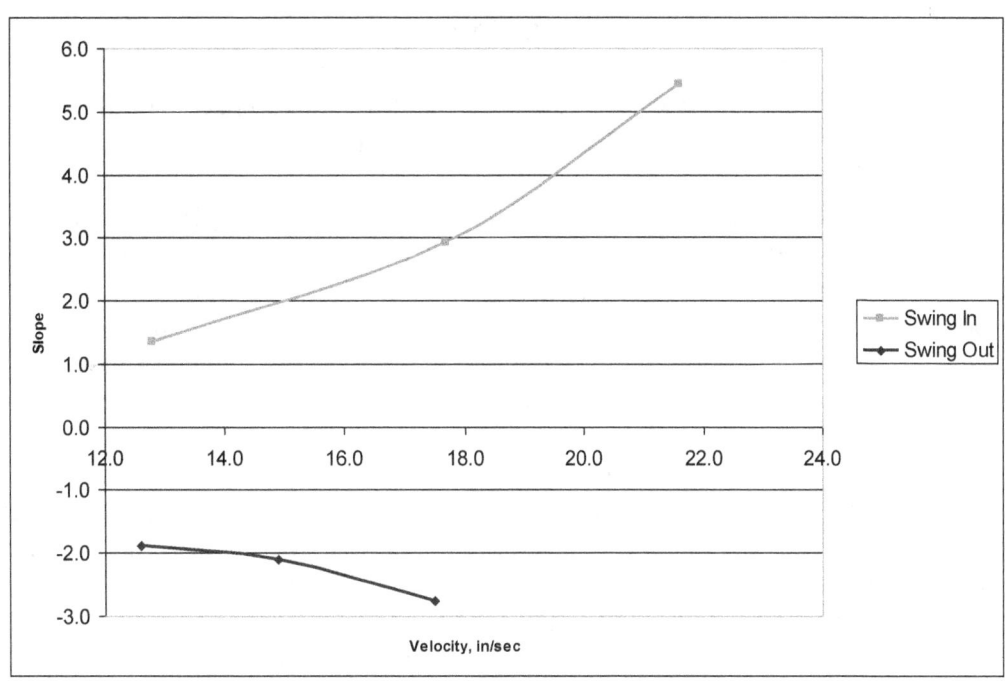

Figure 31.—Mine worker subjects: kneeling 48-in results.

    Obviously, the direction of the swing must be taken into account when deciding whether a positive or negative slope is desired. During boom swing-out, the boom is moving toward the operator, and a positive slope value would indicate the operator is moving faster than the boom. This situation should be a benign case and may indicate the operator can safely interact with a greater boom swing velocity. However, a negative slope value during boom swing-out would indicate the operator-to-boom arm distance is decreasing. This means the boom arm motion is greater than that of the operator. For boom swing-in events, the opposite is true.

    In all postures except one, the trend is quite clear. As boom swing velocity increases, the velocity of the boom arm approaching or receding relative to the operator increases. For a fast swing-out motion, the boom arm approaches the operator faster than he/she can back away, so the boom arm-to-operator distance decreases during the motion. For the fast swing-in motion, the reverse happens. The boom arm moves away from the operator faster than he/she can follow, so the boom arm-to-operator distance increases during the motion. The lone outlier to this trend is the medium velocity data point for the swing-out motion, 72-in standing posture. Results for the 48-in kneeling posture show that for the slowest velocity tested, ~12 in/sec, the operators were unable to keep pace with the moving boom arm for both swing directions. During swing-out motion, the average slope value for the 48-in kneeling posture is always negative, meaning the operator-to-boom arm distance decreased at all velocities.

Discussion

All mine worker study volunteers expressed an inability to keep pace with the fast velocity, as described in Table 3. About 30% commented that the medium velocity was satisfactory to work with for these trials, but felt they would have tired during a longer trial. All of the mine worker volunteers stated they had noticed different operating velocities on equipment of the same model they had used. Most stated that the equipment they were using moved at a rate higher than the slow velocity (the manufacturer's standard) that we used.

The boom arm swinging toward the operator (swing-out) was deemed the more hazardous situation, since the possibility of operator-to-boom arm contact and exposure to pinch point and crush hazards was increased. In some cases, inadvertent control activation may be a contributor to accidents. However, one manufacturer has added a strip stop switch protruding in front of the valves to provide some protection so that in case the operator contacts the strip, the machine shuts down. During swing-in events, the boom arm is moving away from the operator, forcing him/her to follow it. If the boom arm is moving faster than the operator can follow, the distance between the arm and operator will naturally increase. The increased distance and relative motions should provide a safety cushion from the hazards described above. Consider the extreme case where the control lever moved away from the operator rapidly enough so that grasp of the control was lost. The hydraulic actuator would become disengaged, the boom arm would stop, and the operator could catch up with it and regain control.

The most hazardous situations depicted in Figures 27–31 are when the slopes are more negative than −0.5 and the boom arm swing was toward the subject (swing-out). This meant that the distance between the subject and the boom arm was decreasing rapidly and subject-to-boom arm contact could occur. At this slope, the closing velocity between the boom arm and the operator is 50% of the arm's velocity. The target fast swing-out velocity was 16 in/sec. Due to inaccuracies in setting this velocity during human subject testing, the actual average velocity was about 17.5 in/sec. Therefore, the closing velocity between operator and boom arm would be about 9 in/sec. This velocity is similar to the limiting velocities reported by Etherton [1987], OSHA [1987], Karwowski et al. [1992], and DOE [1993].

All tests conducted with the boom swing-out velocity set to the target of 16 in/sec had slopes equal to or less than −0.5. Thus, the rate of 16.0 in/sec for swing-out seems to be too fast for all of the postures tested. The medium rate tested for swing-out averaged 14.7 in/sec and resulted in satisfactory slope values for all postures tested except for the 48-in kneeling posture. However, caution must be exercised because swing velocities at this rate could induce operator fatigue over time.

CONCLUSIONS

Recent MSHA statistics indicate that roof bolter operators are being injured by contact with the boom arm during the bolting cycle. In this study, the subjects were able to keep pace with the manufacturer's standard boom swing rate of 12 in/sec for all of the postures tested except the 48-in kneeling posture. Additional research needs to be conducted for this working height and posture to provide adequate data for conclusions. The data show that subjects are able to keep pace with horizontal boom swing velocities in excess of the manufacturer's standard in some working heights and postures. The data should be viewed with caution because these trials were of a limited duration and the subjects were permitted rest periods as needed. Operator fatigue might be a significant factor in interacting with the higher velocity settings detailed in this research. In addition, other environmental and human factors that were not feasible or within the parameters of this laboratory study, such as floor conditions and operator fatigue may affect safe working velocities.

The hypothesis of this project was that boom arm horizontal swing velocity is an important factor in determining operator safety from pinch point and crush hazards during the boom positioning phase of the bolting sequence. We believe that the data and results support this hypothesis.

ACKNOWLEDGMENTS

We thank the following engineering technicians with the NIOSH Office of Mine Safety and Health Research, Pittsburgh, PA: George F. Fischer (retired) for his efforts in constructing the bolter arm mockup; and Albert H. Cook, Mary Ellen Nelson, and William E. Rossi (retired) for assisting with experiment setup, data collection, and human subject testing. We also express our appreciation to Dean H. Ambrose (retired) and John R. Bartels for helping with mine worker subject scheduling and testing.

REFERENCES

Ambrose DH, Bartels JR, Kwitowski AJ, Helinski RF, Gallagher S, McWilliams LJ, Battenhouse TR Jr. [2005]. Mining roof bolting machine safety: a study of the drill boom vertical velocity. Pittsburgh, PA: U.S. Department of Health and Human Services, Centers for Disease Control and Prevention, National Institute for Occupational Safety and Health, DHHS (NIOSH) Publication No. 2005–128, IC 9477.

Bartels JR, Kwitowski AJ, Ambrose DH [2003]. Verification of a roof bolter simulation model. Warrendale, PA: Society of Automotive Engineers, Inc., technical paper 2003-01-2217.

DOE [1993]. OSH technical reference manual. Chapter 1: Industrial robots. Washington, DC: U.S. Department of Energy, Office of the Assistant Secretary for Environment, Safety, and Health, p. IR-3.

Etherton J [1987]. System considerations of robot end effector speed as a risk factor during robot maintenance. In: Proceedings of the Eighth International System Safety Conference (New Orleans, LA). Unionville, VA: System Safety Society, pp. 434–437.

Karwowski W, Parsaei HR, Soundararajan A, Pongpatanasuegsa N [1992]. Estimation of safe distance from the robot arm as a guide for limiting slow speed of robot motions. In: Proceedings of the 36th Annual Meeting of the Human Factors and Ergonomics Society. Santa Monica, CA: Human Factors and Ergonomics Society, pp. 992–996.

Klishis MJ, Althous RC, Layne LA, Lies GM [1993a]. A manual for improving safety in roof bolting. Morgantown, WV: West Virginia University, Mining Extension Service.

Klishis MJ, Althous RC, Stobbe TJ, Plummer RW, Grayson RL, Layne LA, et al. [1993b]. Coal mine injury analysis: a model for reduction through training. Vol. 8. Accident risk during the roof bolting cycle: analysis of problems and potential solutions. Morgantown, WV; West Virginia University, Mining Extension Service.

McDowell MA, Fryar CD, Hirsch R, Ogden CL [2005]. Anthropometric reference data for children and adults: U.S. population, 1999–2002. Advance Data From Vital and Health Statistics, No. 361, July 7, 2005. Hyattsville, MD: U.S. Department of Health and Human

Services, Centers for Disease Control and Prevention, National Center for Health Statistics. Available at: http://www.cdc.gov/nchs/data/ad/ad361.pdf

MSHA [1994]. Coal mine safety and health roof bolting machine committee: report of findings, July 8, 1994. Arlington, VA: U.S. Department of Labor, Mine Safety and Health Administration, Coal Mine Safety and Health, Safety Division, pp. 1–28.

OSHA [1987]. Guidelines for robotic safety. OSHA directive STD 01-12-002 (instruction pub 8-1.3). Washington DC: U.S. Department of Labor, Occupational Safety and Health Administration, Office of Science and Technology Assessment, September 21.

Turin FC, et al. [1995]. Human factors analysis of roof bolting hazards in underground coal mines. Pittsburgh, PA; U.S. Department of the Interior, Bureau of Mines, RI 9568. NTIS No. PB95-274411.

APPENDIX.—MINE WORKER DATA

This appendix contains tables and figures for each of the mine worker subject tests. Custom software extracted the data from the motion capture files. The analysis was completed by importing the information into Excel. There are five table and figure pairs, each corresponding to a working height/posture combination. Comparison of the actual velocities in the tables below and the target velocities listed in Table 3 shows the variance that occurred during the testing. The slope value represents the rate of change of distance between the operator and boom arm as described in the "Results" section under "Data Analysis: Mine Worker Subjects." The figures are an X-Y scatter plot of the data contained in the associated table.

Table A-1.—Results for standing posture, 72-in height

Subject	Normal velocity				Medium velocity				Fast velocity			
	Swing-out		Swing-in		Swing-out		Swing-in		Swing-out		Swing-in	
	Velocity, in/sec	Slope	Velocity, in/sec	Slope	Velocity, in/sec	Slope	Velocity, in/sec	Slope	Velocity, in/sec	Slope	Velocity, in/sec	Slope
1	12.61	−2.91	12.95	2.22	14.85	−1.49	18.91	2.36	17.65	−2.30	22.56	1.75
2	13.53	−1.31	13.20	−0.88	15.10	−3.19	18.95	2.33	17.79	−1.84	23.48	1.16
3	13.19	−1.91	12.95	0.70	16.86	1.96	17.25	−0.92	17.37	0.85	23.05	2.77
4	12.75	2.05	13.64	−0.10	15.03	0.50	19.43	1.30	17.36	−0.55	22.59	5.42
5	11.89	0.77	12.96	−1.58	15.33	3.02	16.60	−1.88	17.55	−1.91	20.97	0.73
6	12.34	−2.45	12.60	1.01	14.33	−3.13	17.36	2.48	17.27	−3.36	20.82	2.56
7	12.40	−2.41	13.04	0.93	14.65	0.54	17.59	−1.21	17.79	−0.82	21.55	−2.37
8	12.27	−1.01	12.20	0.45	14.02	1.03	17.90	1.27	17.78	−0.12	21.90	2.86
9	12.40	−1.11	12.21	0.30	14.55	0.21	17.14	−0.13	16.67	−1.20	21.07	−0.09
10	12.91	0.18	12.85	−0.05	14.94	0.16	18.37	0.76	18.20	−0.07	20.22	−0.07
11	12.40	1.60	12.83	−1.95	14.84	0.15	16.48	−0.47	17.79	−1.24	22.04	2.29
12	12.57	−0.54	12.17	−0.50	14.25	1.07	16.42	−0.51	16.65	0.92	19.07	−1.14

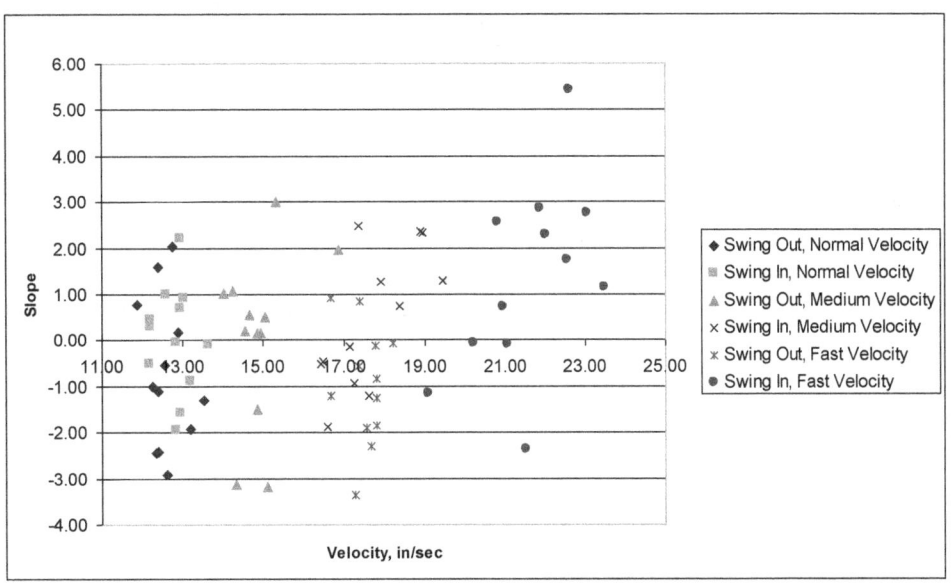

Figure A-1.—Actual mine worker data for standing posture, 72-in height.

Table A-2.—Results for stooping posture, 60-in height

Subject	Normal velocity				Medium velocity				Fast velocity			
	Swing-out		Swing-in		Swing-out		Swing-in		Swing-out		Swing-in	
	Velocity, in/sec	Slope	Velocity, in/sec	Slope	Velocity, in/sec	Slope	Velocity, in/sec	Slope	Velocity, in/sec	Slope	Velocity, in/sec	Slope
1	12.61	-2.53	12.95	1.93	14.85	-1.31	18.91	0.46	17.65	-1.49	22.56	2.27
2	13.53	-0.97	13.20	-0.83	15.10	-1.46	18.95	-1.43	17.79	-2.56	23.48	3.80
3	13.19	-0.40	12.95	0.03	16.86	0.58	17.25	0.18	17.37	1.20	23.05	1.81
4	12.75	2.09	13.64	-0.52	15.03	1.26	19.43	-0.87	17.36	0.15	22.59	2.41
5	11.89	-0.01	12.96	-1.27	15.33	0.37	16.60	-1.24	17.55	-0.83	20.97	1.44
6	12.34	-2.34	12.60	1.31	14.33	-2.60	17.36	1.70	17.27	-3.35	20.82	3.33
7	12.40	0.92	13.04	-1.01	14.65	0.50	17.59	-1.62	17.79	-0.02	21.55	-1.19
8	12.27	0.67	12.20	-1.19	14.02	-0.03	17.90	0.43	17.78	0.63	21.90	1.74
9	12.40	-0.32	12.21	0.37	14.55	-0.19	17.14	0.29	16.67	-1.00	21.07	0.92
10	12.91	0.50	12.85	-0.10	14.94	0.43	18.37	0.51	18.20	0.77	20.22	1.71
11	12.40	1.82	12.83	-1.80	14.84	-1.56	16.48	0.91	17.79	-1.88	22.04	3.13
12	12.57	-1.75	12.17	-0.20	14.25	1.19	16.42	-0.30	16.65	0.91	19.07	-0.68

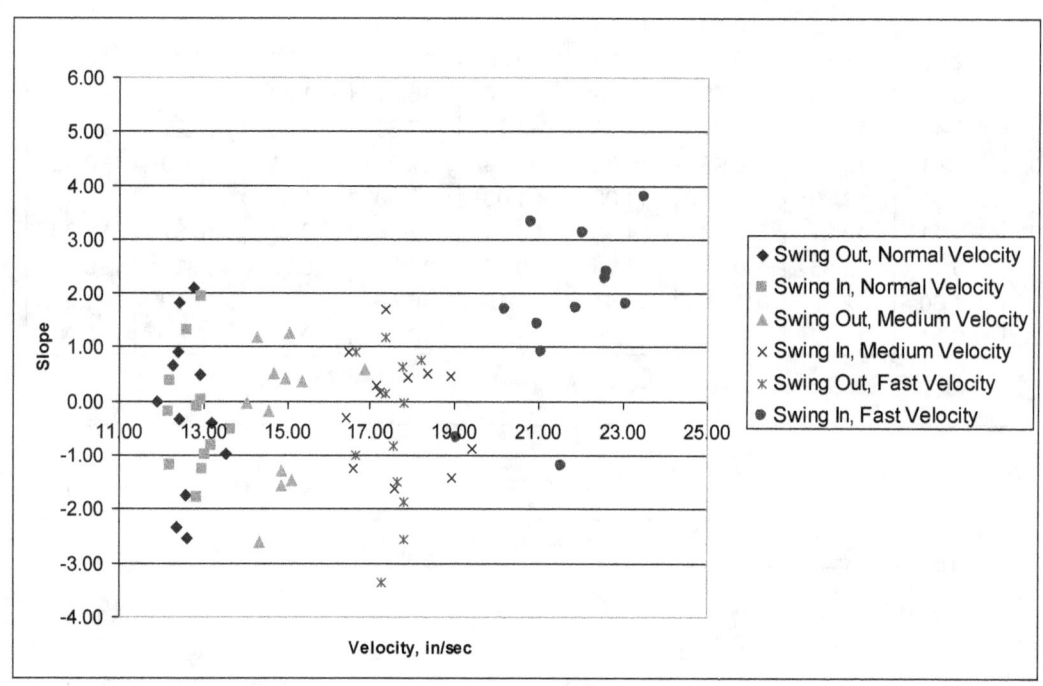

Figure A-2.—Actual mine worker data for stooping posture, 60-in height.

Table A-3.—Results for squatting posture, 60-in height

Subject	Normal velocity				Medium velocity				Fast velocity			
	Swing-out		Swing-in		Swing-out		Swing-in		Swing-out		Swing-in	
	Velocity, in/sec	Slope	Velocity, in/sec	Slope	Velocity, in/sec	Slope	Velocity, in/sec	Slope	Velocity, in/sec	Slope	Velocity, in/sec	Slope
1	12.61	−2.16	12.95	1.07	14.85	−0.57	18.91	0.21	17.65	−1.84	22.56	3.40
2	13.53	−1.59	13.20	−0.02	15.10	−2.58	18.95	−2.32	17.79	−1.71	23.48	−0.25
3	13.19	0.11	12.95	0.07	16.86	0.05	17.25	0.37	17.37	−0.03	23.05	2.58
4	12.75	0.91	13.64	0.05	15.03	0.02	19.43	1.22	17.36	0.63	22.59	1.86
5	11.89	−0.78	12.96	−0.33	15.33	−0.35	16.60	−0.15	17.55	−0.87	20.97	0.77
6	12.34	−1.69	12.60	0.72	14.33	−1.63	17.36	2.29	17.27	−2.48	20.82	2.14
7	12.40	1.16	13.04	−1.19	14.65	−0.13	17.59	−1.62	17.79	−0.05	21.55	−0.83
8	12.27	0.49	12.20	−0.55	14.02	0.28	17.90	1.02	17.78	−0.74	21.90	3.43
9	12.40	0.38	12.21	−0.23	14.55	0.28	17.14	−0.66	16.67	−0.41	21.07	−0.48
10	12.91	1.28	12.85	−0.17	14.94	−0.06	18.37	−0.05	18.20	−0.04	20.22	−0.32
11	12.40	2.30	12.83	−1.87	14.84	−1.08	16.48	1.60	17.79	−1.03	22.04	1.87
12	12.57	−0.15	12.17	−0.91	14.25	−0.54	16.42	−0.48	16.65	0.12	19.07	−0.09

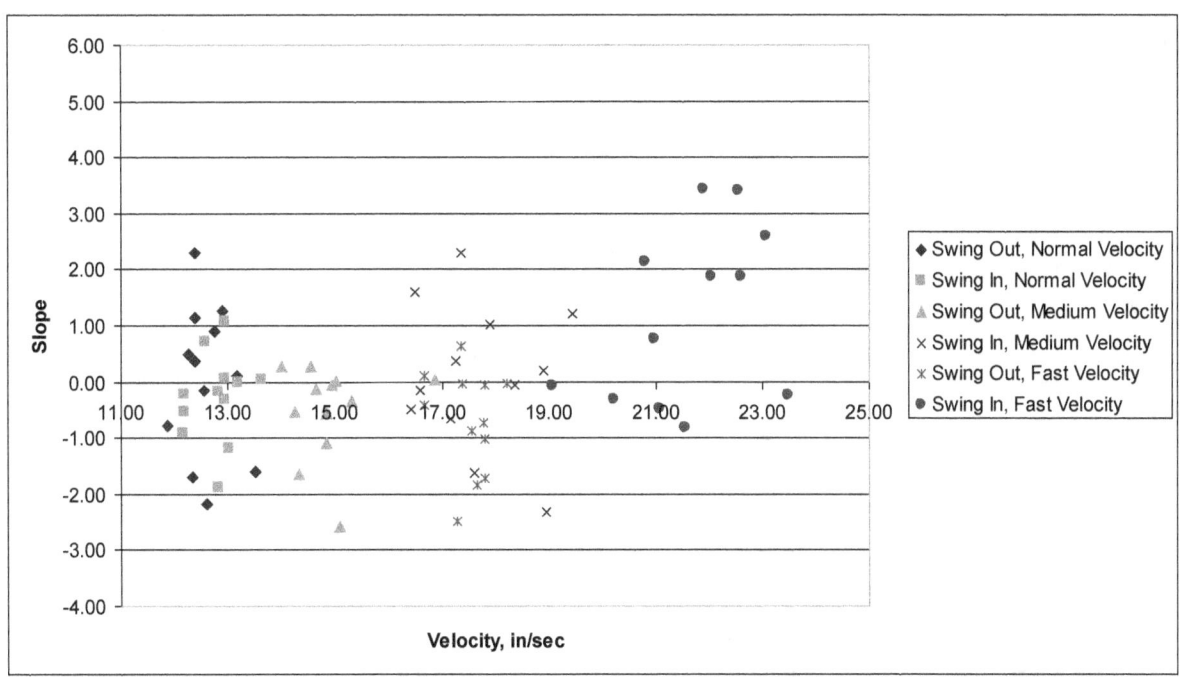

Figure A-3.—Actual mine worker data for squatting posture, 60-in height.

Table A-4.—Results for squatting posture, 48-in height

Subject	Normal velocity				Medium velocity				Fast velocity			
	Swing-out		Swing-in		Swing-out		Swing-in		Swing-out		Swing-in	
	Velocity, in/sec	Slope	Velocity, in/sec	Slope	Velocity, in/sec	Slope	Velocity, in/sec	Slope	Velocity, in/sec	Slope	Velocity, in/sec	Slope
1	12.61	−1.18	12.95	0.02	14.85	−0.68	18.91	−1.59	17.65	−0.72	22.56	1.17
2	13.53	−2.14	13.20	0.42	15.10	−2.86	18.95	2.78	17.79	−3.52	23.48	4.71
3	13.19	0.45	12.95	−0.53	16.86	−0.42	17.25	0.80	17.37	0.24	23.05	0.46
4	12.75	0.28	13.64	−0.70	15.03	0.82	19.43	0.87	17.36	0.24	22.59	1.72
5	11.89	−0.33	12.96	−0.43	15.33	−0.74	16.60	−0.35	17.55	−1.59	20.97	0.94
6	12.34	−1.93	12.60	1.62	14.33	−3.25	17.36	3.19	17.27	−4.24	20.82	4.91
7	12.40	1.36	13.04	−1.31	14.65	1.05	17.59	−1.63	17.79	1.55	21.55	−1.51
8	12.27	0.26	12.20	−0.69	14.02	1.28	17.90	0.83	17.78	−1.12	21.90	1.76
9	12.40	0.07	12.21	−0.65	14.55	−0.99	17.14	0.20	16.67	−1.26	21.07	1.13
10	12.91	1.89	12.85	−0.68	14.94	0.33	18.37	1.15	18.20	0.18	20.22	1.88
11	12.40	0.89	12.83	−1.95	14.84	2.12	16.48	−1.24	17.79	−0.34	22.04	2.61
12	12.57	0.59	12.17	−1.45	14.25	−0.02	16.42	−1.33	16.65	0.88	19.07	−0.65

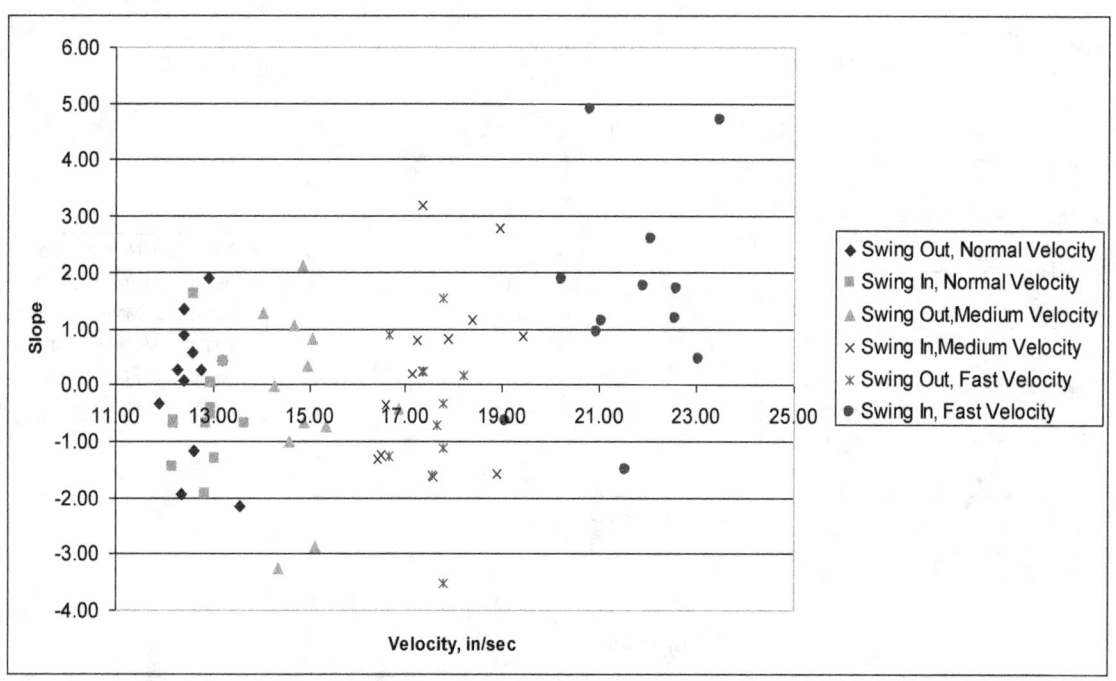

Figure A-4.—Actual mine worker data for squatting posture, 48-in height.

Table A-5.—Results for kneeling posture, 48-in height

Subject	Normal velocity				Medium velocity				Fast velocity			
	Swing-out		Swing-in		Swing-out		Swing-in		Swing-out		Swing-in	
	Velocity, in/sec	Slope	Velocity, in/sec	Slope	Velocity, in/sec	Slope	Velocity, in/sec	Slope	Velocity, in/sec	Slope	Velocity, in/sec	Slope
1	12.61	−3.38	12.95	2.21	14.85	−2.93	18.91	3.54	17.65	−2.34	22.56	6.37
2	13.53	−4.40	13.20	1.72	15.10	−4.92	18.95	4.96	17.79	−7.25	23.48	11.23
3	13.19	−2.02	12.95	1.95	16.86	−2.26	17.25	2.83	17.37	−2.55	23.05	5.83
4	12.75	−1.61	13.64	4.03	15.03	−1.41	19.43	7.25	17.36	−0.98	22.59	9.73
5	11.89	−0.01	12.96	0.32	15.33	−0.37	16.60	1.07	17.55	−0.72	20.97	4.06
6	12.34	−3.36	12.60	2.00	14.33	−4.06	17.36	2.96	17.27	−6.46	20.82	4.27
7	12.40	−1.86	13.04	−0.74	14.65	−2.92	17.59	−0.58	17.79	−4.90	21.55	1.66
8	12.27	−1.59	12.20	1.09	14.02	−0.50	17.90	5.60	17.78	−1.10	21.90	8.75
9	12.40	−0.43	12.21	0.59	14.55	−0.22	17.14	0.20	16.67	−1.43	21.07	3.73
10	12.91	1.37	12.85	0.42	14.94	−0.30	18.37	2.04	18.20	0.97	20.22	2.19
11	12.40	−2.89	12.83	2.38	14.84	−2.92	16.48	3.20	17.79	−2.74	22.04	5.27
12	12.57	−2.57	12.17	0.07	14.25	−2.51	16.42	1.85	16.65	−3.49	19.07	2.20

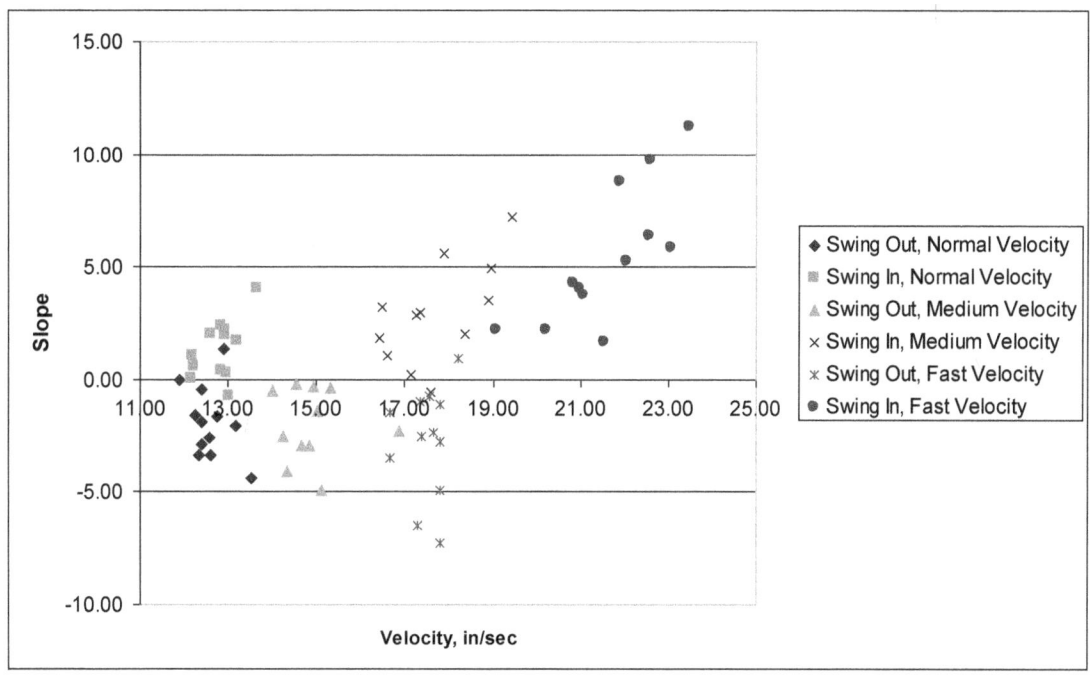

Figure A-5.—Actual mine worker data for kneeling posture, 48-in height.

www.ingramcontent.com/pod-product-compliance
Lightning Source LLC
Chambersburg PA
CBHW081803170526
45167CB00008B/3311